ADVANCED MATHEMATICAL TOOLS IN METROLOGY III

Series on Advances in Mathematics for Applied Sciences – Vol. 45

ADVANCED MATHEMATICAL TOOLS IN METROLOGY III

Editors

P Ciarlini
CNR – Istituto di Applicazione del Calcolo, Roma, Italy

M G Cox
National Physical Laboratory, Teddington, UK

F Pavese
CNR – Istituto di Metrologia, Torino, Italy

D Richter
Physikalisch-Technische Bundesanstalt, Berlin, Germany

World Scientific
Singapore • New Jersey • London • Hong Kong

Published by

World Scientific Publishing Co. Pte. Ltd.

P O Box 128, Farrer Road, Singapore 912805

USA office: Suite 1B, 1060 Main Street, River Edge, NJ 07661

UK office: 57 Shelton Street, Covent Garden, London WC2H 9HE

Library of Congress Cataloging-in-Publication Data

Advanced mathematical tools in metrology III : proceedings of the Euroconference : Berlin, Germany, 25–28 September 1996 / editors, P. Ciarlini . . . [et al.].

p. cm. -- (Series on advances in mathematics for applied sciences ; vol. 45)

Proceedings of the Third International Workshop on Advanced Mathematical Tools in Metrology.

Includes bibliographical references and index.

ISBN 9810229186 (alk. paper)

1. Mensuration -- Congresses. 2. Physical measurements -- Congresses. I. Ciarlini, P. II. International Workshop on Advanced Mathematical Tools in Metrology (3rd: 1996 : Berlin, Germany) III. Series

QA465.A283 1997

530.8'01'51--dc21 97-18874

CIP

British Library Cataloguing-in-Publication Data

A catalogue record for this book is available from the British Library.

This book is printed on acid-free paper.

Printed in Singapore by Uto-Print

Foreword

This volume collects selected, revised contributions at the Third International Workshop on Advanced Mathematical Tools in Metrology held in Berlin, Germany, in September 1996. As in the volumes of the two preceding workshops, the papers in this volume present and promote the application of mathematical and in some extent computational methods to metrology.

To a great extent, advances in metrology depend on the clever use of mathematical and computational tools. The workshop aimed at increasing the extent of cooperation between metrologists, mathematicians and software people. Emphasis was placed on the benefits that mathematical and software tools can bring to problems facing researchers and practitioners of metrology. To limit the width of topics under discussion, the workshop had been focused from the methodological point of view on

- uncertainty problems, stochastic methods;
- inverse problems, numerical methods;
- software and algorithm validation;

and from the metrological point of view on

- geometric and manufacturing measuring technique;
- physical measurements in medicine.

Repeating a well-tried principle, the contributions are divided into tutorial papers, accepted papers and short communications representing selected poster presentations. The first ones have an overview or introductory character, and the second and the third ones report on concrete activities.

Berlin, February 1997 The Editors

EUROCONFERENCE "ADVANCED MATHEMATICAL TOOLS IN METROLOGY"

Coordinator: F.Pavese, Istituto di Metrologia "G.Colonnetti" , Torino, IT
(EU Grant ERBCHECCT 940254 to IMGC)
Berlin Chairman: D.Richter, Physikalisch-Technische Bundesanstalt, Berlin, DE

INTERNATIONAL SCIENTIFIC COMMITTEE

P.Ciarlini, Istituto per le Applicazioni del Calcolo "M.Picone", Roma, IT
M.G.Cox, National Physical Laboratory (NPL), Teddington, UK
R.S.Davis, Bureau International des Poids et Mesures, Sèvres
H.Kunzmann, Physikalisch-Technische Bundesanstalt (PTB), Braunschweig, DE
F.Pavese, Istituto di Metrologia "G.Colonnetti" (IMGC), Torino, IIT
D.Richter, Physikalisch-Technische Bundesanstalt (PTB), Berlin, DE

ORGANISED BY

CNR, Istituto di Metrologia "G.Colonnetti", (IMGC) Torino
Physikalisch-Technische Bundesanstalt (PTB), Berlin, Germany

LOCAL COMMITTEE

D.Richter, Physikalisch-Technische Bundesanstalt (PTB), Berlin, DE
H.Möeck, Physikalisch-Technische Bundesanstalt (PTB), Berlin, DE

Sponsored by
European Communities (Human and Capital and Mobility Programme)), Consiglio Nazionale delle Ricerche (IT), Physikalisch-Technische Bundesanstalt (DE), ECMI, SIMAI (IT), MEMC (IT), DPG (DE), Visual Numerics (DE), JF Lehmanns Fachbuch-handlung (DE), Helmholtz Fonds e.v. (DE)

These Proceedings were made possible by a grant of
MEMC Electronic Materials SpA (Italy)

For the first two Volumes see this Series vol. 16 (1994) and vol.40 (1996)

א

Proceedings Editors
Patrizia Ciarlini (Istituto per le Applicazioni del Calcolo, CNR, Roma, Italy), **Maurice G.Cox** (National Physical Laboratory, Teddington, UK), , **Franco Pavese** (Istituto di Metrologia "G.Colonnetti", CNR, Torino, Italy), **Dieter Richter** (Physikalisch-Technische Bundesanstalt, Berlin, Germany)

Table of Contents

Full Papers

Short Communications

ADVANCED MATHEMATICAL TOOLS IN METROLOGY III

Advanced Mathematical Tools in Metrology III
Edited by P. Ciarlini, M.G. Cox, F. Pavese & D. Richter

THE ISO *GUIDE TO THE EXPRESSION OF UNCERTAINTY IN MEASUREMENT*: A BRIDGE BETWEEN STATISTICS AND METROLOGY

W. BICH
CNR - IMGC, Istituto di Metrologia «G. Colonnetti», str. delle cacce, 73, 10135 Torino, Italy

This paper overviews the ISO *Guide to the Expression of Uncertainty in Measurement*, considering its impact in the old *querelle* about random and systematic errors. Introduces a generalisation of the *Guide* to multivariate cases, showing the role of covariances and analyses a number of difficulties in the construction of intervals of confidence, although approximate.

1 Introduction

Few topics have raised in the metrological environment so passionate a discussion for such a long time as the evaluation of measurement uncertainty. Indeed, few topics are as pervasive as this, which involves virtually any conceivable measurement activity. The reference document is, since 1993, the *Guide to the Expression of Uncertainty in Measurement*[1], published by ISO in the name of an impressive list of international organizations. The *Guide* is a milestone in the treatment of measurement uncertainty. It states precise rules for assigning the uncertainty in a wide range of experimental situations. There is nothing really new in it, but a rather conspicuous transfer of knowledge between different scientific subjects. In this sense, the *Guide* is a bridge between metrology and statistics, and, as it happens in such cases, it is somewhat criticised for its difficulty by metrologists, and for its crudity by statisticians. This note should be intended as an attempt of a metrologist to walk from his side of the bridge to the opposite end, and as a description of the difficulties and problems encountered on his way.

2 The measurement model

It is assumed that the quantity Y to be measured, *the measurand*, is constant during measurement. This is more a simplifying than a mandatory condition.

It is assumed that n other quantities X_i are observed, on which the measurand depends according to a function[a]

$$Y = f\left(X_1, X_2, \dots X_n\right) \; . \tag{1}$$

[a] It is to be noted that the analytical form of the function may not be known explicitly.

This scheme, sometimes called the *indirect measurement*, encompasses most of the experimental situations. For example, even the simple case of direct reading of an instrument is described in terms of Eq. 1. The output reading X depends on the value of the measurand Y at the input of the instrument and on the transfer characteristic of the instrument according to a functional relationship $X = g(Y)$, so that, inverting it, one may write

$$Y = f(X,S) = \frac{X}{S} \quad , \tag{2}$$

in which S, the *sensitivity*, is defined as the reciprocal of the slope of $X = g(Y)$.

The observed quantities X_i, i.e., the independent variables in Eq. 1, are the *input quantities*, whereas the measurand Y is often called the *output quantity*.

So far, everything is deterministic. It is assumed that each X_i is estimated by a corresponding *input estimate* x_i, which is considered a random quantity, related to X_i by

$$x_i = X_i + \varepsilon_i \tag{3}$$

under the condition

$$E(x_i) = X_i \tag{4}$$

which can be written explicitly

$$E(\varepsilon_i) = 0 \quad , \tag{4a}$$

where $E(x)$ denotes expectation of the random variable x. In this paper the equivalent notation $\overline{x}$ will also be used.

Therefore, the random outcome of an observation is considered as the sum of the constant value of the quantity and an unknown random error with zero expectation. The input estimates have to be *unbiased*.

The y value resulting from a set of input estimates x_i,

$$y = f(x_1, x_2, \ldots x_n) \quad , \tag{5}$$

is the *output estimate*. It is obviously a random quantity. A first-order Taylor expansion of Eq. 5 about $f(\overline{x_1}, \overline{x_2}, \ldots \overline{x_n})$, i.e., about the expectations of the observations, gives

$$y = f(x_1, x_2, \ldots x_N) \approx f(\overline{x_1}, \overline{x_2}, \ldots \overline{x_n}) + \sum_{i=1}^{n} \left. \frac{\partial f}{\partial x_i} \right|_{x_i = \overline{x_i}} (x_i - \overline{x_i}) \quad , \tag{6}$$

from which, by taking expectations in both sides, and by virtue of conditions 4a, one obtains

$$\overline{y} \equiv E[f(x_1, x_2, \ldots x_n)] \approx f(\overline{x_1}, \overline{x_2}, \ldots \overline{x_n}) \ , \tag{7}$$

that is to say

$$E(y) \approx Y \quad . \tag{8}$$

Therefore, in the current model of measurement, the output estimate y is (to the first order) an unbiased estimate of the measurand Y under the condition that input estimates x_i are unbiased.

3 Uncertainty of measurement, one measurand

The basic idea of the *Guide* is to identify the uncertainty of measurement of a quantity with the standard deviation[b] of the estimate of that quantity. This is by no means a new idea, as the well-known error theory, dating back to Gauss, is based on the very same idea. What is new is its implementation, as we will discuss later on.

Equation 6, remembering Eq. 8, can be written as

$$y - \overline{y} \approx \sum_{i=1}^{n} \left. \frac{\partial f}{\partial x_i} \right|_{x_i = \overline{x_i}} (x_i - \overline{x_i}) \ , \tag{9}$$

which is the error propagation law.

By squaring both sides of Eq. 9, and by taking expectations, one obtains

$$\sigma_y^2 \approx \sum_{i=1}^{n} \left. \frac{\partial f}{\partial x_i} \right|_{x_i = \overline{x_i}}^2 \sigma_{x_i}^2 + 2 \sum_{i=1}^{n-1} \sum_{j=i+1}^{n} \left. \frac{\partial f}{\partial x_i} \frac{\partial f}{\partial x_j} \right|_{x_{i,j} = \overline{x}_{i,j}} \mathrm{cov}(x_i, x_j) \quad , \tag{10}$$

---

[b] Standard deviation $\sigma(x)$ is defined ad the positive square root of the variance $V(x) = E\left[(x - \overline{x})^2\right]$

or, by rearranging the products in a more compact form,

$$\sigma_y^2 \approx \sum_{i=1}^{n}\sum_{j=1}^{n} \frac{\partial f}{\partial x_i}\frac{\partial f}{\partial x_j}\bigg|_{x_{i,j}=\bar{x}_{i,j}} \mathrm{cov}\left(x_i, x_j\right) \quad , \tag{11}$$

in which $\mathrm{cov}\left(x_i, x_i\right) = \sigma_{x_i}^2$.

Equations 10 and 11 are the uncertainty propagation law.

4 The multivariate case

The *Guide* does not treat explicitly the multivariate case, although good treatments can be found in the literature[2, 3], and even in a national standard [4]. The multivariate case occurs when m measurands are obtained from a common set of n input variables

$$\begin{array}{l} y_1 = f_1\left(x_1, x_2, \ldots x_n\right) \\ y_2 = f_2\left(x_1, x_2, \ldots x_n\right) \\ \ldots\ldots\ldots \\ y_m = f_m\left(x_1, x_2, \ldots x_n\right) \end{array} \quad , \tag{12}$$

or, in matrix notation

$$\boldsymbol{y}_{(\mathrm{mx1})} = \boldsymbol{f}\left(\boldsymbol{x}_{(\mathrm{nx1})}\right) \quad . \tag{13}$$

This is a very common situation in metrology, since a large fraction of the calibration certificates issued by calibration laboratories refer to a set of instruments or standards. Examples are the calibrations of sets of weights, of gauge blocks, of resistors, etc. by means of the same experimental set-up. The writing of a model in terms of vectors and matrices, such as that of Eq. 13, may be difficult. For an example, see ref. 3.

Here, the basic condition 4a is

$$E(\boldsymbol{\varepsilon}) = \boldsymbol{0} \, . \tag{14}$$

In the same way in which the second moment about the mean (the variance) represents the dispersion of a random scalar quantity, dispersion of a random vector $\boldsymbol{u}$ is expressed by its variance-covariance matrix $\boldsymbol{\psi}_{\boldsymbol{u}}$, here called for brevity *covariance*

$$\boldsymbol{\psi}_{\boldsymbol{u}} = E\left[\left(\boldsymbol{u} - \bar{\boldsymbol{u}}\right)\left(\boldsymbol{u} - \bar{\boldsymbol{u}}\right)^T\right] \quad . \tag{15}$$

The matrix element $\psi_{u_{i,j}} = E\left[\left(u_i - \overline{u_i}\right)\left(u_j - \overline{u_j}\right)\right]$ is

$$\psi_{u_{i,j}} = \begin{cases} V(u_i) & j=i \\ \operatorname{cov}(u_i, u_j) & j \neq i \end{cases} \quad . \tag{16}$$

Consequently, the covariance matrix is arranged in such a way that its diagonal elements are the variances of the components of the random vector and the off-diagonal elements are their covariances. Declaration of the covariance of measurands is recommended in the *Guide* for the multivariate case, without further discussion. As an alternative, the dimensionless *correlation matrix* ρ_u may be specified, which is related to covariance by

$$\rho_{u_{i,j}} = \frac{\psi_{u_{i,j}}}{\left[V(u_i)V(u_j)\right]^{1/2}} \tag{17}$$

It is easily verified that, by virtue of the Cauchy-Schwartz inequality,

$$-1 \leq \rho_{u_{i,j}} \leq +1 . \tag{18}$$

The matrix ψ_y, the covariance of the output vector $\boldsymbol{y}$, can be defined in terms of the covariance of the input ψ_x by a straightforward extension of the scalar propagation law.
By virtue of Eq. 14, it is easy to check that

$$\psi_x = E(\varepsilon_x \varepsilon_x^T) = \psi_{\varepsilon_x} \quad , \tag{19}$$

that is, the covariance of $\boldsymbol{x}$ is simply that of its errors.
A first-order Taylor expansion of Eq. 13 about $E(\boldsymbol{x})$ gives

$$\boldsymbol{y} \approx \overline{\boldsymbol{y}} + \boldsymbol{J}_x \boldsymbol{\varepsilon}_x , \tag{20}$$

where

$$\boldsymbol{J}_x \equiv \left\{ \frac{\partial y_i}{\partial x_j} \right\} \tag{21}$$

is the well-known Jacobian, evaluated at $\boldsymbol{x} = \overline{\boldsymbol{x}}$. Equation 20 can be written

$$\boldsymbol{\varepsilon}_y \approx \boldsymbol{J}_x \boldsymbol{\varepsilon}_x \quad . \tag{22}$$

By postmultiplying both sides of Eq. 22 by $\varepsilon_y^T \approx \left(\boldsymbol{J}_x \boldsymbol{\varepsilon}_x\right)^T$, and by taking expectations, one has simply

$$\psi_y \approx J_x \psi_x J_x^T \quad . \tag{23}$$

Equation 23 is the fundamental law of uncertainty propagation. If y is a linear function of x, $y = Lx$, where L is a linear operator, such as, for instance, the least-squares estimator, then

$$\psi_y = L\psi_x L^T \tag{24}$$

is an exact law.

The matrix element $\psi_{y_{h,k}}$ in Eq. 24 has the form

$$\psi_{y_{h,k}} = \sum_{i=1}^{n}\sum_{j=1}^{n} \frac{\partial y_k}{\partial x_i}\frac{\partial y_h}{\partial x_j}\psi_{x_{i,j}} \quad . \tag{25}$$

With $h = k$, Eq. 25 reduces to Eq. 10, that is, to the scalar uncertainty-propagation law. If ψ_x is diagonal, that is, if the input quantities are uncorrelated, Eq. 25 becomes

$$\psi_{y_{h,k}} = \sum_{i=1}^{N} \frac{\partial y_h}{\partial x_i}\frac{\partial y_k}{\partial x_i}\sigma^2(x_i) \quad , \tag{26}$$

which shows that, even with uncorrelated input estimates, correlations may arise between output estimates. This fact is well known, but the metrologist is usually reluctant to believe that *his* particular field needs multivariate treatment. I will sketch here a simple example, showing the influence of the reference standard[5, 6], in the attempt to shake our consolidated habits.

5 The correlation due to traceability

Let for simplicity assume that two standards of unknown values y_i are calibrated by comparison to one reference standard of value R. The model 13 has here the linear form

$$y = Lx, \qquad L = \begin{bmatrix} -1 & 0 & 1 \\ 0 & -1 & 1 \end{bmatrix}, \qquad x = \begin{bmatrix} x_1 \\ x_2 \\ R \end{bmatrix} , \tag{27}$$

in which x_i are the differences observed between the reference standard and each unknown. Let also

$$\psi_x = \begin{bmatrix} \sigma_C^2 & 0 & 0 \\ 0 & \sigma_C^2 & 0 \\ 0 & 0 & \sigma_R^2 \end{bmatrix}, \tag{28}$$

that is, the observations are supposed equal-variance and uncorrelated. Application of Eq. 24 yields

$$\psi_y = \begin{bmatrix} \sigma_C^2 + \sigma_R^2 & \sigma_R^2 \\ \sigma_R^2 & \sigma_C^2 + \sigma_R^2 \end{bmatrix}, \tag{29}$$

from which

$$\rho_{y_{1,2}} = 1 - \sigma_C^2 / \left(\sigma_C^2 + \sigma_R^2\right). \tag{30}$$

This simple example demonstrates the quite obvious fact that the traceability uncertainty is common to both estimates, that is to say, in statistical terms, is their covariance. This basic result holds also for comparison schemes more complicated than that described by Eq. 27, such as those involving least-squares adjustments. Actually, since any calibration of a set of artefact standards, such as weights, may be represented by a vector model such as that in Eq. 27, the presence of covariances between output estimates due to traceability might be much more pervasive than it has been believed so far. It would, for example, be interesting to explore the covariances existing between standards individually calibrated at different times by the same laboratory, or between standards calibrated by a series of one-to-one comparisons with a set of correlated reference standards.

Often, the traceability contribution to uncertainty σ_R is dominant with respect to comparison repeatability σ_C, so that $\rho_{y_{1,2}} \approx 1$. This is, for example, typical for mass standards.

6 Comments

Equation 10, and the more general Eq. 23, express the law of propagation of variances and covariances. They require that input estimates may be modelled by Eqs. 3 and 13, respectively, subject to conditions 4, viz. 14. These equations are the pillars upon which it is possible to establish a relationship between the natural world and the numbers representing the observations of it. They make it possible to express the first two moments of output distribution in terms of the corresponding moments of input distributions without any assumption concerning the distributions themselves, except the condition that errors have zero mean.

This whole framework has been, and still is, questioned in that it could not encompass systematic errors, such as those coming from a bias in an instrument, or from an imperfect correction, or from an imperfect model. These errors, by their very nature, cannot have zero mean. The concept itself of mean can be meaningless for some of these errors. Statistics can only be applied to random quantities. These, from a frequency-based probability viewpoint, may result from tossing a die, or extracting numbers from a box, or repeatedly reading an instrument. In other words, a population must exist behind the sample, may it be a sample of dimension one. Within this traditional framework, the estimated value of an external standard or of a physical constant, or a correction based on uncertain or subjective data are definitely non-random quantities, since there is no population behind them. Therefore their errors have a generally non-zero, although unknown, value, and the uncertainty associated with these values cannot be interpreted in a statistical way. As a consequence, in this scheme the estimate of the measurand is generally biased, and the evaluation of its uncertainty should take account of its bias. This point of view has traditionally led to separate treatments of uncertainties referring to random and systematic quantities and, sometimes, to a final combination of these two contributions. The price to be paid for such a rigid frequency-based interpretation of probability is rather high. As uncertainty is no longer a purely statistical quantity, statistical tools cannot be used as they stand, but need heavy modifications[7, 8].

The solution adopted in the *Guide* for this serious problem lies in a broader interpretation of randomness: random is whatever cannot exactly be known, or predicted, and the probability of an event quantifies the degree of belief in its occurrence. In this scheme, probability distributions model the available amount of information, be it coming from experiment or from other sources, including subjective judgement. A certain, deterministic datum is a random quantity having a Dirac δ distribution. The value of π is now known to a little over 10^8 digits, the remainder is randomness[9]. The advantage of this point of view, with respect to the traditional one, is that all uncertainty components, as they have the same statistical nature, can be treated statistically.

7 The estimate of the measurand

The adoption of such a point of view has important consequences not only on the uncertainty of an estimate, but also on the estimate itself. In fact, the requirement of zero-mean errors, necessary for a correct propagation of uncertainties, has to be re-interpreted in subjective terms. The estimate is the "best" estimate not in absolute terms, but simply within the knowledge of the experimenter. Accordingly, since the knowledge of the experimenter is usually based not only on "objective" data, coming from the observation of the physical world, but also on "subjective" judgement, such as that sometimes adopted for evaluating a systematic effect, these two components should both contribute to the estimate. A good example is the so-called cosine error, that is, the error arising, for example, from misalignment of a reference length-measuring device with respect to the length to be measured. This results in an overestimation, whose correction is frequently based on subjective evaluation. The need for such procedures is far from being fully realized in the metrological community. Most of us are reluctant to correct an estimate for an effect which is judged to influence it, but about which objective data are not available. It is often preferred to include the subjective evaluation as an uncertainty component. This practice should be avoided.

A further source of criticism is connected with the co-existence in the *Guide* of two viewpoints of probability which are to some extent conflicting, in the sense that they may lead in specific cases to different results[10]. Attempts have been made to harmonise the two viewpoints by using the Maximum Entropy Principle[11], but the question seems to deserve further investigation.

8 The estimation of the input covariance

The input covariance ψ_x in Eq. 23 has to be estimated, and its estimate is used for estimating the output covariance ψ_y. If a sample is available for an input component, the experimental variance s^2 of the sample is adopted. In the negative, other empirical, or prior, knowledge or even subjective judgement has to be used. In the language of the *Guide*, the two methods of evaluation are referred to as Type A and Type B, respectively.

The main difficulty concerning Type A evaluations lies in that seldom observations are a random sample taken from stationary processes. Most phenomena are, to different extent, time-dependent, so that they can hardly be interpreted in this scheme, as, for example, the sample variance may diverge with the sample size. In such cases, preliminary data filtering is necessary in order to obtain a meaningful measure of dispersion. These filtering techniques range from a simple subtraction of drift, obtained in different ways[12], to the well-known high-pass filtering leading to Allan variance[13]. My feeling is that the present model for observations, based essentially on stationary, white processes, is not sufficiently general. In addition, time series and signal analysis techniques ought to be applied to a wider range of experimental situations.

A concern which is common to both types of uncertainty evaluation regards the reliability of the estimated ψ_x. This is specially relevant in least-squares adjustments, in which the weighting matrix is the inverse of ψ_x. Recursive techniques for harmonising initial and posterior knowledge about ψ_x have been developed[14, 15], and the extension of their use should be considered.

9 The construction of a confidence region

The *Guide* is primarily concerned with the estimation of the output variance, and the problem of constructing a confidence interval having a specified probability level is only addressed in a specific Annex. The terms *level/interval of confidence* are introduced, as an alternative to the traditional ones, probably as a consequence of discrepancies within the group of experts responsible for the development of the *Guide*.

The output standard deviation has a reliability which results from the reliability of each individual input variance, according to the classical Welch-Satterthwaite formula[16, 17]

$$\nu_{\text{eff}} = s^4(y) \Bigg/ \sum_{i=1}^{n} \frac{\left(\dfrac{\partial y}{\partial x_i} s_{x_i}\right)^4}{\nu_i}\,, \tag{31}$$

in which ν_i is the degrees of freedom of the i-th input variance and ν_{eff} is the *effective degrees of freedom* of the output variance. The formula holds only if the output estimate y is t-distributed, that is, if the Central Limit Theorem can be applied. Although the *Guide* is reassuring on the wide range of experimental situations for which the Theorem is valid, the suspect is that these are less than expected. In particular, correlation between input quantities could play an important role.

There is a second concern about the degrees of freedom of a subjectively estimated input variance. This is a somewhat new concept, whose (subjective) evaluation yields invariably very high values which in some cases tend to dominate in Eq. 31, thus making the effective degrees of freedom of the output largely insensitive to the actual number of observations.

In the general multivariate case, the analytical difficulties inherent in the construction of a confidence region in the space of the output, could be overcome by using computer-intensive methods such as Montecarlo simulation or bootstrap resampling[18, 19].

10 Conclusions

As a problem is usually more interesting than a smooth situation, focus in this article has been more on the former than on the latter. This might give the false impression that the *Guide* is far from being a well-established document. The *Guide to the Expression of Uncertainty in Measurement* gives to the metrological community, for the first time, precise recipes for evaluating uncertainties in measurements. Therefore its usefulness is invaluable. We never should forget that an uncertainty is seldom required to be known to better than some percent...

References

1. ISO, *Guide to the Expression of Uncertainty in Measurement*, Switzerland, First ed. (1993).
2. A. Allisy, in: *Proceedings of course LXVIII: Metrology and Fundamental Constants*, International School of Physics *Enrico Fermi*. eds. A. Ferro Milone *et al.* (North-Holland, Amsterdam, 1980).
3. W. Bich et *al.*, *Metrologia* **30**, 495 (1993/1994).
4. DIN 1319 Teil 4 (1985).
5. W. Bich, *Metrologia* **27**, 111 (1990).
6. W. Bich, *Metrologia* **33**, 181 (1996).
7. M. Grabe, *Metrologia* **23**, 213 (1986/1987).
8. M. Grabe, *Phys. Lett. A* **165**, 124 (1992).
9. S. James Press, *Bayesian Statistics: Principles, Models, and Applications* (Wiley, New York, 1989).
10. P M Lee, *Bayesian Statistics: An introduction* (Edward Arnold, London, 1989).
11. K. Weise, W. Wöger, *Meas. Sci. Technol.*, **3**, 1 (1992).
12. C. M. Sutton, M. T. Clarkson, *Metrologia* **30**, 487 (1993/94).
13. D. W. Allan, *IEEE Trans.* **IM-36**, 646 (1987).
14. E. R. Cohen, in: *Precision Measurement and Fundamental Constants II*, B. N. Taylor, W. D. Phillips, Eds. Natl. Bur. Stand. (U.S.) Spec. Publ. **617**, 391 (1984)
15. E. R. Cohen, *IEEE Trans.* **IM-44**, 475 (1995).
16. B. L. Welch, *Biometrika* **34**, 28 (1947).
17. F. E. Satterthwaite, *Biometrics Bull.* **2**(6), 110 (1946).
18. B. F. J. Manly: *Randomization and Montecarlo Methods in Biology* (Chapman and Hall, London, 1991).
19. E. W. Noreen: *Computer Intensive Methods for Testing Hypotheses* (Wiley, New York, 1989).

Advanced Mathematical Tools in Metrology III
Edited by P. Ciarlini, M.G. Cox, F. Pavese & D. Richter

BOOTSTRAP ALGORITHMS AND APPLICATIONS

PATRIZIA CIARLINI

CNR, Istituto per le Applicazioni del Calcolo "M. Picone",
V.le del Policlinico 137, 00161 Roma, Italy

Bootstrap methodology allows to compute accuracy measures of statistical estimators with minimal model assumptions in a totally automatic way. Bootstrap algorithms can easily be implemented on a computer and used also by researchers, not expert in statistics. Simple and more advanced bootstrap applications, such as in linear and non-linear regression, diagnostics, time series analysis, are addressed with some examples on real data.

1. Introduction

Bootstrap is a constructive methodology developed in 1979 by Efron[6,7] to estimate or make statistical inference in most real problems by a direct computation from data sets. It does not require theoretical calculations and can give results also with mathematically complicated estimators: it is alternative to analytic solutions or to approximation methods, that are derived from a postulated probability model for the data, since only in a few cases explicit formula are known.

Bootstrap belongs to the class of resampling methods, such as the jackknife, commonly used in biomedical applications, or the permutation method, that can be easily performed when the sample size is small. For an easy computation of the bootstrap estimators, simulation approach based on the Monte Carlo method can simply and easily be implemented also on a PC. However, as with other statistical tools, attention must be paid to fulfill the hypoteses on which they are based (see Lepage and Billard[13]).

Bootstrap technique received and continues to receive an increasing attention not only from statisticians and mathematicians but also from applied researchers because it is:

— easy-to-use, since it does not require any knowledge on probability distribution of the sample, but employes resampling from the empirical distribution function,

— useful in experimental studies when it is impossible to repeat the experiment a sufficiently large number of times to identify the unknown probability function underlying the data, because of the high costs or of the change in the experimental conditions,

— automatic, because it avoids complex formulae.

This paper briefly introduces the bootstrap technique and the most commonly used algorithms. In particular, it is aimed to stimulate to use bootstrap in metrology, an interdisciplinary area of research, which provides precision data and estimates of their uncertainties, that may depend on a very large number of parameters both of stochastic and of deterministic nature. The ability of bootstrap to indicate accuracy estimates in metrological applications was firstly introduced in Ciarlini and Pavese[2]).

Specific algorithms to compute confidence intervals are discussed in section 3. Section 4 illustrates some applications of bootstrap to simple statistics. Successively more complicated bootstrap applications are outlined: linear and non-linear regression, with extension to the problem of the diagnostic of unusual values, time series analysis, where the case of dependent data will be addressed. The adopted numerical techniques and the resampling will be shown up, to draw attention to the correct definition of the algorithms, since every experimental condition must be satisfied.

Real data used in the examples are taken from Efron and Tibshirani[10] or were obtained from Istituto di Metrologia "G.Colonnetti", CNR, Turin. Every computation is obtained by using the software described in Ciarlini and Regoliosi[3]. A general theory for the bootstrap is illustrated in Shao and Tu[14], while a more direct and practical exposition is given in Efron and Tibshirani[10].

2. Bootstrap algorithm

The bootstrap is a computer-based method to estimate the accuracy of an estimator. The basic idea is to mimic what has been observed in the *Real World* (the unknown probability distribution F and n observed data $\mathbf{x} = (x_1, ..., x_n)$ —the original sample) by simulating a *Bootstrap World* (an estimated probability model $\hat{F}$ and a bootstrap sample $\mathbf{x}^* = (x_1^*, ..., x_n^*)$). In this way the probability distribution of any statistic of interest $R = R(\mathbf{x}, F)$ is approximated by the bootstrap distribution constructed by using B replications $R_b^* = R(\mathbf{x}_b^*, \hat{F})$, b = 1,..., B, one for each bootstrap sample; $\hat{F}$ is the empirical distribution function that puts probability mass 1/n on each of the observed point x_j.

An explicit description of the bootstrap technique to measure the accuracy of $R = R(\mathbf{x}, F)$ (i.e. standard error or bias) is reported in the following Monte-Carlo algorithm, where the resampling operation is performed by means of a random number generator:

(0) for the *original sample* $(x_1, x_2, ..., x_n)$ compute $\hat{R} = R(\mathbf{x}, \hat{F})$

(1) draw, randomly with replacement, a *bootstrap sample* $\mathbf{x}_b^*$ of size n from $\hat{F}$: this implies that $x_j^* = x_i$ for any index j, i = 1, ..., n.,

(2) for each bootstrap sample $\mathbf{x}_b^*$ calculate the *bootstrap variable* $R_b^* = R(\mathbf{x}_b^*, \hat{F})$

Repeat steps 1 - 2 B times independently to get $R_b^* = R(\mathbf{x}_b^*, \hat{F})$, b= 1, ..., B

(3) the sample distribution of R is approximated by the *bootstrap distribution* of R^* so defined:

$$H^*(r) = \text{Prob}_{\hat{F}}\,(R^* \leq r) \qquad (2.1)$$

Finally, estimate a bootstrap accuracy measure, such as bias or standard deviation, for the given statistics:

$$\text{bias}^*_B = \bar{R}^* - \hat{R} \qquad \text{where} \quad \bar{R}^* = \frac{1}{B}\sum_{b=1}^{B} R^*_b$$

$$s^*_B = \left[\frac{1}{B-1}\sum_{b=1}^{B}\left(R^*_b - \bar{R}^*\right)^2\right]^{1/2}$$

Bootstrap technique can be used both in parametric and non parametric contexts. In the first case parametric model $F(\theta)$ is assumed for the data, and the sample **x** is used to estimate the parameter $\hat{\theta}$ (one or more). Then B bootstrap samples are drawn from $F(\hat{\theta})$ instead of $\hat{F}$ in step 2. Non parametric bootstrap may fail when the empirical distribution is not a good estimate of the true F and this may occur especially in the extreme tails (importance sampling and smoothed bootstrap estimators are suggested in Shao and Tu[14]).

The computational time for the above iterative algorithm depends on the value of B and on the complexity of the computation of R^* in steps 2. The theory defines $B \rightarrow \infty$ as the ideal choice, but a suitable value of B must be fixed. It is known that a bootstrap estimate of standard error usually has relatively little bias. But a good behaviour has to be assured for each computed bootstrap estimate, whatever the statistics or the sample size n can be. In practice recommended values for the bias or the standard error are $B \in [50, 100]$ and for confidence interval or hypothesis tests are $B \in [1000, 2000]$. More sophisticated choices can be defined by using the coefficient of variation of the statistics, i.e. the ratio of the standard deviation s^*_B to the expectation.

3. Bootstrap Confidence Intervals

The three most used bootstrap algorithms to construct confidence intervals at level $100(1\text{-}2\alpha)\%$ — here simply called $(1\text{-}2\alpha)$CI— are here briefly discussed (see also Efron[8,9]). For example they provide a valid alternative to the construction of standard $(1\text{-}2\alpha)$CI using the traditional normal approximation, since they give more accurate values when the Normal assumption is not so realistic to describe the given data.

The percentile method is the simplest one, being directly applicable from data. It is invariant under variable transformation, but it is first order accurate[1]

[1] • A confidence interval I_n is said to be k-th order asymtotically accurate if $P\{R \in I_n\} - (1-\alpha) = O(n^{-k/2})$

only. It is considered by Efron as "a computational algorithm for extending the range of effectiveness of the standard intervals", where the correct variable transformation is implicitly applied.

The boostrap estimate of the bootstrap distribution in Eq.(2.1) is computed by using the replications:

$$H^*_B(r) = \frac{\sum_{b=1}^{B} I(R^*_b \le r)}{B} \tag{3.1}$$

where $I(R^*_b \le r)$ is 1 if the relation in brackets is true, zero otherwise. The algorithm to get the percentile values consists in the following steps:

(1) draw B bootstrap samples $\mathbf{x}^*_b$ from $\hat{F}$ and compute R^*_b;

(2) put in an increasing order the list of values R^*_b, b=1, ..., B;

The percentile bootstrap (1-2α)CI is $[R^{*(\alpha)}_B, R^{*(1-\alpha)}_B]$, where $R^{*(\alpha)}_B = H^{*-1}_B(\alpha)$ is 100·α-th percentile of the bootstrap distribution and corresponds to the value in the 100·α-th position of the ordered list constructed in step (2).

The second method is the bootstrap-*t* percentile, which uses a pivotal transformation of the bootstrap variable R^*:

$$T^* = \sqrt{n}\,\frac{R^* - \hat{R}}{s^*} \tag{3.2}$$

where $\hat{R}$ is assumed to be a good estimate of the true R. Now the percentile related to T^* are estimated by the bootstrap. This method is suitable for location parameters, and reproduces the standard intervals when data are normally distributed, but it is not invariant under transformation of the variable. It has accuracy of the second order, but it needs a good estimate of the standard deviation s* of the statistics. To get this estimate Hinkley[12] suggests to use a second-level bootstrap, then the total cost of this algorithm becomes proportional to P*B, where P is the number of resampling of second level (usually it is sufficient P = 25). The algorithm was also described in Ciarlini et al.[1].

An improved version of the percentile method is the bias corrected and accelerated method BC_a. It uses two parameters, the acceleration and the bias correction, to take into account a possible change in the standard deviation and a possible bias in the approximation of the normal distribution by applying the variable transformation. To estimate the acceleration as in Efron and Tibshirani[10], the *jackknife* estimates of $\hat{R}$ are used to get $\hat{a}$, and the Monte Carlo algorithm is only used to provide the bias correction $\hat{z}_0$.

Consider n samples of size n-1, that can be deduced from **x** by deleting one element, say $\mathbf{x}_{(i)} = (x_1, \dots, x_{i-1}, x_{i+1}, \dots, x_n)$, i = 1,..., n. For each of them compute the jackknife estimates of the statistics $\hat{R}_{(i)} = \hat{R}(\mathbf{x}_{(i)})$ and their average $\hat{R}_{(.)}$, then the acceleration is estimated by:

$$\hat{a} = \frac{\sum_{i=1}^{n}\left(\hat{R}_{(.)} - \hat{R}_{(i)}\right)^3}{6\left[\sum_{i=1}^{n}\left(\hat{R}_{(.)} - \hat{R}_{(i)}\right)^2\right]^{3/2}} \qquad \text{where } \hat{R}_{(.)} = \frac{1}{n}\sum_{i=1}^{n}\hat{R}_{(i)}$$

$\hat{z}_0$ is directly obtained by using the replications R^*_b:

$$\hat{z}_0 = \Phi^{-1}\left(\frac{\#(R^*_b \leq \hat{R})}{B}\right) \quad (\Phi^{-1} \text{ inverse of standard normal cdf})$$

The $(1-2\alpha)BC_a$ confidence interval is given by $[\, R_B^{(\alpha_1)}, R_B^{(\alpha_2)}]$, where

$$\alpha_1 = \Phi\left(\hat{z}_0 + \frac{\hat{z}_0 + z^{(\alpha)}}{1-\hat{a}(\hat{z}_0 + z^{(\alpha)})}\right) \quad \alpha_2 = \Phi\left(\hat{z}_0 + \frac{\hat{z}_0 + z^{(1-\alpha)}}{1-\hat{a}(\hat{z}_0 + z^{(1-\alpha)})}\right)$$

and $z^{(\alpha)}$, is 100α-th percentile point of the standard normal distribution.

Main characteristics of this method are: the need of a good estimate of the acceleration, that is invariant under transformation and it is second order accurate. Other types of intervals can be constructed by taking $\hat{z}_0$ or $\hat{a} = 0$.

4. Simple bootstrap applications

The bootstrap algorithms of the previous sections are now applied to some sample surveys, commonly used in analysing experimental data. In the examples of sections 4.1 and 4.2 the dataset of the lutenizing hormone (Efron and Tibshirani[10] pag.92) is used: it is here named EX1. In every computation B is fixed at 1000 and $(1-2\alpha) = 0.95$.

4.1 Huber M-estimate of the location parameter

Let the statistics R be the Huber estimate of the location parameter as in Davison[5], then the bootstrap variable is $H^* = R(\mathbf{x}^*, \hat{F})$, that is obtained as the solution of the linear equation

$$\sum_{i=1}^{n} \varphi(x_i^* - H^*) = 0$$

where
$$\varphi(y) = \begin{cases} -c & \text{for } y < -c \\ y & \text{for } -c \le y \le c \\ c & \text{for } y > c \end{cases}$$

It is known that for $c \to \infty$ H approximates the mean, while for c = 0 the median. The M-estimate with c = 4 for the original sample in EX1 (last data modified by adding 4) is $\hat{H} = 2.4787$, and its bootstrap estimates of accuracy are:

$bias_B^* = -.00252$ $\qquad s_B^* = .11744$

$(1\text{-}2\alpha)CI_B^*$ (percentile method) $\qquad$ [2.2688, 2.5833]

$(1\text{-}2\alpha)CI_B^*$ (bootstrap-*t* method) $\qquad$ [2.2587, 2.6260]

$(1\text{-}2\alpha)CI_B^*$ (BC_a method) $\qquad$ [2.2750, 2.5938]

The three intervals are compatible, but the computational cost is higher for the bootstrap-*t* method, where a value P = 100 is used in the second level bootstrapping; the standard estimate of H in Eq.(3.2) is computed by considering the deviation from the trimmed mean.

4.2 Median

No exact formula is known for the standard error of the median, while bootstrap estimates can be easily computed. Also the difference between median and mean is a statistical parameter used to decide if the mean can be used instead of the median. If the data analysis is performed by means of bootstrap, the following bootstrap variable is considered:

$$d^* = \text{med}(\mathbf{x}^*) - \bar{x}^*$$

For data in EX1 is $\hat{d} = -.10417$, and the bootstrap estimates of accuracy are:

$bias_B^* = .03758$ $\qquad s_B^* = .05552$

$(1\text{-}2\alpha)CI_B^*$ (percentile method) $\qquad$ [-.16458, .05000]

$(1\text{-}2\alpha)CI_B^*$ (BC_a method) $\qquad$ [-.22083, -.03750]

These results imply that the difference between median and mean should not be neglected, since the more accurate estimate of the CI* by BC_a algorithm does not contain the zero value.

4.3 Skewness and curtosis

In order to have an indication of the shape of the unknown probability distribution of data the following parameters, related to the third and fourth moments of the probability distribution, are to be considered. Simple bootstrap simulations can assess accuracy estimates also in cases where only small size samples are available. The bootstrap variables, that correspond to the skewness and curtosis, are the following:

$$\gamma_1^* = \frac{m_3^*}{s^{*3}} \qquad \gamma_2^* = \frac{m_4^*}{s^{*4}} - 3$$

where m_3^* and m_4^* are the 3-th and 4-th moments of the boostrap distribution, and for example m_3^* is estimated by $m_{3,B}^* = \frac{1}{B}\sum_{b=1}^{B}\left(x_b^* - \bar{x}^*\right)^3$.

The boostrap methodology is applied to data that are generated by means of the Beta(3,1) distribution function. For a sample of size 50 the bootstrap standard deviation, bias and 0.95CI for γ_1 and γ_2 are computed and reported in Table 1. The analysis of the CI* outlines that γ_1 can be considered statistically different from zero, that is to say that its bootstrap distribution is skewed, in agreement with the known shape of Beta(3,1).

Table 1: Bootstrap accuracy measures (B = 1000) for skewness and curtosis.

	original values	stand dev*	bias*	0.95 CI* percentile	BC$_a$
γ_1	0.9042	0.2727	-0.79e-2	0.3915, 1.496	0.4813, 1.567
γ_2	2.741	0.6969	0.47e-1	1.8470, 4.473	1.9570, 5.011

5. Application to regression

Bootstrap can be used to provide accuracy of any estimate which can be required in linear regression problems, when the assumption of Normal distribution function is not assumed to describe the errors behaviour, as in Freedman[11]. In Shao and Tu[14] it is established that bootstrap estimators may be more accurate than those obtained by assuming the normal approximation. An iterative procedure for the selection of the regression model by bootstrapping is given in Ciarlini and Regoliosi[3]. Moreover the bootstrap can provide an alternative to the standard methodology to identify outliers or unusual values in regression. The technique can also be applied in non linear regression problems or when the mathematical solution is not provided by means of the least squares estimator.

5.1 Linear case

The classical linear regression model consists of n data points $\mathbf{x}_1 \ldots \mathbf{x}_n$ where each $\mathbf{x}_i = (\mathbf{c}_i, y_i)$ is a pair of the covariate m-vector $\mathbf{c}$ and of the response y. It is given in matrix form by:

$$\mathbf{y} = \mathbf{C}\,\mathbf{a} + \mathbf{e} \qquad \mathbf{C} \in R^{n\times m} \tag{5.1}$$

where $\mathbf{e}$ is assumed to be a zero-mean random variable of an unknown distribution function F; Eq. (5.1) is still valid for the case of curve regression, where each element of the nxm design matrix $\mathbf{C}$ is now of monomial form.

The probability structure of the linear model is given in Eq. (5.1) and F, say $(\mathbf{a}, F)$. The LSE of the regression parameters $\mathbf{a}$ are obtained as solution of the normal equations and under the equal variance assumption $\sigma_i = \sigma$:

$$\hat{\mathbf{a}} = (\mathbf{C}^T\mathbf{C})^{-1}\mathbf{C}^T\,\mathbf{y}$$

In order to analyse the results of the regression by means of the bootstrap methodology Efron proposed the resampling based on residuals (RB) where $\mathbf{x}$ is assumed to be deterministic and e_i are i.i.d.

Bootstrap data $\mathbf{y}^*$ are simulated (cfr. steps (1) (2) in section 2) by replacing $(\mathbf{a}, F)$ by $(\hat{\mathbf{a}}, \hat{F})$, i.e. putting probability mass 1/n on each $\hat{e}_j$, the probability mechanism gives $e_i^* = \hat{e}_j$, i and j = 1,..., n, where the residuals are $\hat{\mathbf{e}} = \mathbf{y} - \mathbf{C}\,\hat{\mathbf{a}}$:

$$\mathbf{y}_b^* = \mathbf{C}\,\hat{\mathbf{a}} + \mathbf{e}_b^* \qquad b = 1, \ldots, B \tag{5.2}$$

In this linear system the matrix $\mathbf{C}$ does not depend on the bootstrap iteration, hence, as in the algorithm proposed in Ciarlini and Pavese[2], it can be preprocessed before entering into the iteration to get $\mathbf{a}_b^*$ with Eq. (5.2). Algorithms of section 3 can be used to compute bootstrap confidence intervals for the regression parameters or for any other statistics commonly used in data analysis such as Fisher F or the multiple correlation coefficient.

Bootstrap differently applies when $\mathbf{x}$ is random. This approach is called bootstrapping pairs $\mathbf{x}_i = (\mathbf{c}_i, y_i)$ in Efron and Tibshirani[10] (PB). Now bootstrap data are simulated by randomly sampling from a multinomial (m+1)- distribution function; it is equivalent to generate a sample of integers $(i_1, \ldots, i_n)$ to define the index j, that corresponds to a row in $\mathbf{C}$:

$$\mathbf{x}_i^* = (\mathbf{c}_j, y_j) \qquad i, j = 1, \ldots, n \tag{5.3}$$

The matrix $\mathbf{C}$ in Eq.(5.2) now depends on the bootstrap iteration, say $\mathbf{C}_b^*$, and rank deficiency may occur due to the duplication of rows, see Eq. (5.3), in the resampling step. Hence in solving the linear system to get $\mathbf{a}_b^*$ for each b, the generalized inverse must be used.

Due to this construction of the design matrix, the PB method is less sensitive to model assumption. Hence it is robust against heteroschedasticity or possible correlation between data, while RB is more model-dependent, but is more efficient from a computational point of view.

A useful application of the regression analysis is for the problem of diagnostics. As discussed in Cook and Weisberg[4], an alternative method to the residuals analysis is the study of the influence of unusual values on the least squares estimates. This is performed by measuring possible difference in the solutions via the Cook's distance. Data are modified by deletion of case by case and the following n distances have to be compared to identify possible outliers:

$$D_{(k)} = \frac{(\mathbf{a}_{(k)}-\mathbf{a})^T \mathbf{C}^T \mathbf{C} (\mathbf{a}_{(k)}-\mathbf{a})}{(m-1)\, s^2_{(k)}} \qquad k = 1, ..., n \qquad (5.4)$$

where $\mathbf{a}_{(k)}$ are the regression parameters computed after having deleted the k-th data from the original dataset, $s^2_{(k)}$ is the estimated variance. The k-th data is considered influential if $D_{(k)}$ is large, and this may happen also with a small residual. It is possible to define a Cook's distance in the case of deletion of subset of data. Bootstrap confidence intervals for $D_{(k)}$ can be constructed.

In the following example a comparison of the bootstrap estimates for regression parameters obtained by the RB and the PB approaches, is shown, using a subset of the database (23 data series collected by different authors) of specific-heat measurements in the temperature range 273-1400 K, already analysed to identify multiple outliers in Ciarlini et al.[1]. In the present analysis only N = 21 series, for a total number of n = 264 data, are considered, in order to draw the attention on the analysis to do in order to identify what series can be considered outlier (now in Eq.(5.4) k, k = 1,..., N, selects the series number).

A regression cubic polynomial model that contains 20 translational effects, for a number of coefficient m = 24 in Eq.(5.1), is applied: in Table 2 only the four polynomial coefficients are reported in the second column. The bootstrap accuracy measures obtained by using RB algorithm are in columns 3-5. Those obtained by PB algorithm are in columns 6-8. The confidence intervals are estimated by means of the BC_a algorithm with B = 1000 and $1\text{-}2\alpha = 0.95$.

The maximum value of the Cook's distance, reached for series k = 19, is reported in the last row of Table 2. Its bootstrap confidence interval is costructed by means of the percentile algorithm, since its bootstrap distribution appears heavily skewed. The bootstrap estimates of the polynomial parameters in the two approaches are similar: this is probably due to the good choice of the regression model and to the robustness of the least squares estimators. Concerning the accuracy measures of the Cook's distance, the PB approach provides a larger bias and a longer CI*, then the 19-th series can be better identified as an outlier via the PB algorithm.

Table 2: Comparison of RB and PB approaches to polynomial regression of metrological data.

	LSQ values	stand dev*	bias*	CI* (RB method)	stand dev*	bias*	CI* (PB method)
a_0	0.0998	0.0006	0.2e-4	0.0987–0.1009	0.0005	-0.2e-4	0.0989, 0.1010
a_1	0.0143	0.0011	-0.7e-5	0.0121–0.0166	0.0011	-0.1e-3	0.0125, 0.0168
a_2	0.0084	0.0007	-0.2e-5	0.0071–0.0100	0.0007	0.2e-4	0.0068, 0.0096
a_3	0.0023	0.0014	0.3e-4	-0.0005–0.0051	0.0015	0.3e-3	-0.0010, 0.0048
$D_{(19)}$	0.2247	0.1019	-0.140	0.0007–0.3583	0.328	0.119	0.0009, 1.1820

5.2 Non linear case

The RB can be also applied to a non linear model, depending on m parameters a_j , $y_i= f(x_i, \mathbf{a}) + e_i$. It is an extension of the model in Eq.(5.1) and in the same error assumptions a LS estimator is generally considered. In order to simulate bootstrap data:

$$y^*_i = f(x_i , \hat{\mathbf{a}}) + e^*_i \qquad i = 1, ..., n \qquad (5.5)$$

an iterative computation is required to solve the minimum problem associated to Eq.(5.5), i.e. a Newton-type algorithm to get the LS estimates $\hat{\mathbf{a}}$. Moreover the same iterative algorithm should be used to compute $\mathbf{a}^*_b$, b= 1, ..., B, for which the minimum is reached in the following espression:

$$\sum_{i=1}^{n} (y^*_{i,b} - f(x_i, \hat{\mathbf{a}}))^2$$

When the computation of the given statistics involves iterations, the bootstrap algorithm may become very time expensive. Instead of solving the minimum problem, the following approximation to $\mathbf{a}^*_b$, called one-step bootstrap, is suggested in Shao and Tu[14]:

$$\tilde{\mathbf{a}}^*_b = \hat{\mathbf{a}} + \mathbf{M}^{-1}(\hat{\mathbf{a}}) \sum_{i=1}^{n} e^*_i \nabla f(x_i, \hat{\mathbf{a}}) \qquad (5.6)$$

It avoids solving iteratively B minimum problems, since both the **M** matrix and the partial derivatives indicated in $\nabla f(x_i, \hat{\mathbf{a}})$ have already been computed to get the LS $\hat{\mathbf{a}}$. Then only by bootstrapping the residuals and evaluating $\tilde{\mathbf{a}}^*_b$, the bootstrap accuracy measures of the non linear regression parameters can be easily computed (cfr. step (3) in section 2).

6. Autoregressive or moving average time series

It has to be emphasised that appropriate bootstrap procedure must be applied to mimic the true mechanism, that generates the original data, if reliable bootstrap estimates have to be provided. When time series are processed, the previous algorithms, developed for i.i.d. data are no more applicable, since the observations are correlated. Two main strategies to apply bootstrapping to dependent data are available: if the structure of the dependency is known[2], the resampling of moving block data can be performed, otherwise the resampling must be applied from properly defined residuals, to take into account the model structure, in analogy with the RB method in the previous section.

In the first approach the n original data are divided in groups or blocks of fixed size K, n = Km, and the resampling is applied to the m groups, considered as primary units, to maintain the dependency in the bootstrap data. Then the consistency of the estimates obtained by resampling of moving block of data strictly depends on how exactly is known the size K, and this may be a difficult choice.

The resampling method based on residuals is more dependent on the model assumptions, however it can be automatically used without adding difficult analytic computations.

Let us now consider this latter method to derive the bootstrap distribution for the parameters of an autoregressive model of order 1:

$$y_t = \mu + \theta\, y_{t-1} + e_t \qquad t = 0, \pm 1, \pm 2, ..$$

where $\mu = 0$ is assumed for simplicity and e_t are i.i.d zero-mean random variables with variance σ^2. Data $(y_1, ..., y_n)$ are observed with the initial observation y_0 and used to compute in the least squares sense the autoregressive parameter and the residuals:

$$\hat{\theta} = \frac{\sum_{t=1}^{n} y_t y_{t-1}}{\sum_{t=1}^{n} y_t^2} \qquad (6.1)$$

$$r_t = y_t - \hat{\theta}\, y_{t-1}$$

From the centered residuals $r_t - \bar{r}$, t = 1, ..., n a resample e_1^* ..., e_n^* is taken, and the bootstrap estimates for the parameter are obtained by solving:

---

[2] • A sequence of r.v. $\{X_t, t = 0, \pm 1, \pm 2, .\}$ is m-dependent if there is a integer $m \geq 0$ such that $\forall$ t $\{...X_{t-1}, X_t\}$and $\{X_{t+1+m}, X_{t+m+2}, ...\}$ are independent

$$e_0^* = y_0$$
$$y_t^* = \hat{\theta}\, y_{t-1}^* + e_t^* \qquad t = 1,\ldots, n$$

where y_0 is taken as initial bootstrap observation. The bootstrap distribution for θ^* is estimated using the LS estimates θ_t^*, in Eq.(3.1).

The AR(1) model is here applied to EX1 with $\mu \neq 0$, then the residuals must not be centered before boostrapping. In Table 3 the bootstrap accuracy measures for the model parameters and the variance are shown.

Table 3: Bootstrap accuracy measures (B = 1000) for AR(1) model applied to lutenizing hormone data.

	LSE	stand dev*	bias*	.95 CI* Percentile	Bias Corrected
μ	2.4103	0.5764	0.4479	1.957, 4.248	1.212, 2.933
θ	0.5876	0.1347	0.0014	0.320, 0.861	0.310, 0.848
s^2	0.2087	0.3248	0.3147	0.139, 1.291	0.081, 0.537

Acknowledgements

The author is indebted to Dr. Giuseppe Regoliosi, co-authors on many works on the bootstrap technique on which this review paper is based.

References

1. P.Ciarlini *et al*, in *Advanced Mathematical Tools in Metrology*, eds. P.Ciarlini, M.G.Cox, R.Monaco and F.Pavese, (World Scientific, Singapore, 1994)
2. P.Ciarlini an d F.Pavese, *Numerical Algorithms* **1-4**, 479 (1993)
3. P. Ciarlini and G.Regoliosi, *Il software ABOOS per le applicazioni di bootstrap non parametrico: Parte I*, (Quad. IAC **7**, 1996)
4. R.Cook and S.Weisberg, *Residuals and Influence in Regression*, Monograph on Statistics and Applied Probability, (Chapman and Hall, London, 1982)
5. A.C.Davison, in *Supplement to Handbook of Applicable Mathematics* (eds. W. Ledermann, E. Lloyd, S. Vajda and C. Alexander, John Wiley, N. Y., 1988)
6. B.Efron, *Ann. of Stat.* **7**, 1 (1979)
7. B.Efron, *SIAM Review* **21**, 460 (1979)
8. B.Efron, *JASA* **82**, 171 (1987)
9. B.Efron and R.Tibshirani, *Statistical Science* **1**, 54 (1986)
10. B.Efron B. and R.Tibshirani, *An Introduction to the Bootstrap*, (Monograph on Statistics and Applied Probability 57, Chapman e Hall, Londra, 1993)
11. D. A.Freedman, *Ann. of Stat.* **9**, 1218 (1981)
12. D.Hinkley, *J. R. Statist. Soc. B* **50**, 321 (1988)
13. R.Lepage and L.Billiard eds., *Exploring the limits of bootstrap*, (John Wiley, New York, 1992)
14. J.Shao and D.Tu, *The Jackknife and Bootstrap*, (Springer Verlag, New York, 1995)

Advanced Mathematical Tools in Metrology III
Edited by P. Ciarlini, M.G. Cox, F. Pavese & D. Richter

THE TTRSs : 13 ORIENTED CONSTRAINTS FOR DIMENSIONING, TOLERANCING & INSPECTION

A. CLEMENT
Dassault Systèmes,
9 Quai Marcel Dassault, B.P. 310, 92156 SURESNES Cedex, FRANCE
email : aclement@dassault-systemes.e-mail.com

C. VALADE , A. RIVIERE
Institut Supérieur des Matériaux et de la Construction Mécanique,
3 Rue Fernand Hainaut, 93407 SAINT-OUEN Cedex, FRANCE
email : 101346.2215@compuserve.com

The state of art of CAD-CAM model during the 90's deals with feature, parametrics and constraint based models. In that direction, an integrated model using 13 signed constraints for dimensioning, tolerancing and metrology in CAD-CAM systems is proposed : the TTRS (Topologically and Technologically Related Surfaces) model. The power of this model is constituted by three main elements : the TTRS which is an object made of pair of surfaces, the MGRS which is the Minimum Geometrical Reference Set of those surfaces and the 13 signed constraints. A constraint is a mathematical relationship between variable elements. If those elements are geometrical elements, such as MGRSs, then the constraints are called *relative positioning signed constraints*.

Any scientific law has to be able to independently express itself from a Cartesian coordinate system, by handling intrinsic mathematical objects : scalars, groups, rings, bodies, vectors, tensors. From an historic point of view, that is particularly true for the most ancient of any science, which is the Euclidean geometry.

Around 1981, two events appeared apparently independent the one from the other, but linked to the previous note, and, which both strongly marked technology evolution of the mechanical production.

The first of those two events was the birth of the first three dimensional CAD-CAM system. That system was representing all the geometrical objects using their analytic expressions in a Cartesian coordinate system without never using intrinsic mathematical objects.

The second of those two events was the ASME Y14.5 standard publication regarding to machined part tolerancing. That document marked the convergence to a long and empirical process of parametrics definition of actual machined parts. Noteworthy achievement : that document never has recourse to the slightest analytic expression, even if the notion of coordinate system is omnipresent within that document.

The limit of use of those two approaches soon appeared, excluding themselves and mutually from an always more advanced technological world. On one hand, the emergence of computerized measuring machines has required the use of Cartesian coordinate system for metrological computations.

On the other hand, the necessity of a convivial approach for the CAD-CAM software's has constrained their designers to use parametric definitions based on distance and angle notions. Solely the systematic alliance of the parametric approach and of the analytic approach will allow to solve the problem of the Man- Machine relationship. That report aims to propose a strict mathematical representation of the parametric approach based on the theory of solid displacement group which will be entirely compatible with the analytic approach.

Through several useful examples, the applying of that theory will be shown in a single frame and regarding to the formalism of problems dealing with dimensioning, tolerancing and dimensional inspection of machined parts.

1. The TTRS model : A mathematical Analysis

A 3-dimensional orthogonal normed Cartesian coordinate system is a very powerful mathematical object, endowed with a complex structure. Such a coordinate system is composed with the Euclidean affine space E^3, which is associated with the normed vectorial space $\overrightarrow{E^3}$. That combination, due to René Descartes, of those two spaces allows either the analytical representation of any solid, or to follow the course of any deformation of continuous medium.

The parametrics is a process by which it is defined distances and angles just necessary and sufficient, in order to reconstruct a geometrical object composed with a finite set of solids.

The goal is not here the description of the continuous medium description but is solely the relative position of lines, surfaces and volumes preserving their form or aspect between two changes of position. The general structure is therefore too much powerful.

A simplified mathematical structure has to be determined. That structure shall be into use for man, who just knows distance and angle notion, but that structure shall be compatible with the computer, which is just able to manipulate analytical representations.

That goal is obtained as explained in the following section. It shall be essential to construct Cartesian coordinate system. For that, the absolutely necessary affine space E^3 is held, however solely a part of the vectorial space is held. The vectorial space $\overrightarrow{E^3}$ is composed with the additive group structure of the real numbers and with the scalar multiplication.

That vectorial structure furnishes on one hand, the affine space orientation and on the other hand, the capacity to associate the Euclidean norm. Concretely , that vectorial structure is represented using the three unitary vectors $\vec{i}$, $\vec{j}$, $\vec{k}$.

When the displacements are restricted to solids, it is possible to restrict the mathematical structure to the solid displacement group. Generally, the associated vectorial space has a dimension smaller than 3.

From that time onwards, solely solids will be handled. At the time when one will wish to make deformations of continuous mediums, one will leave such created tools, but on the other hand the obtained Cartesian coordinate system will be compatible with those deformations and analytical representations.

1.1 Class of surfaces

It is proved that the displacement group of the solids in the 3-dimensional Euclidean space is subdivided in 11 subgroups. The noteworthy property of those subgroups is as follows.

When any displacement is applied to a surface, that displacement may always be decomposed in two parts :

- the first part which effectively displaces the surface (and which can be called significant displacement),
- the second part (possibly the identity displacement) which leaves the surface invariant in simply making it sliding on itself.

On the other hand, it is proved that all displacements, which leave a surface invariant, belong to the same subgroup. That subgroup is called the invariance subgroup of the surface. Then reciprocally, it is possible to feature all the surfaces using the subgroups, which leave them invariant.

It is proved that there exist 7 subgroups for which a feature can be constructed (see table 1).

1.2 Definition of a Minimum Geometric Reference Set (MGRS)

For each subgroup of the displacement group {D}, the class of globally invariant surfaces was highlighted where appropriate. In each class there are minimum geometric elements such as points, lines, planes or helices, representing the invariant Cartesian system element and featuring the positional geometric parameters of any surfaces. Those invariant geometric elements will be systematically used in order to construct a Cartesian coordinate system. They will be referred to as MGRSs, standing for Minimum Geometric Reference Sets. As seen on Table 1, one can defined MGRS for each of seven classes of surfaces.

	Dimension	Corresponding subgroup of {D}	MGRS (Situation Feature)
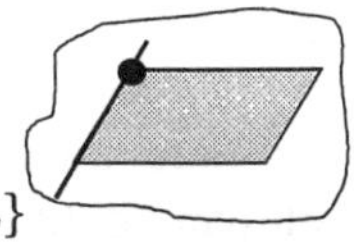 {E}	0	{E} Identity displacement	Any plane of *E3* Any line of *E3* Any point of *E3*
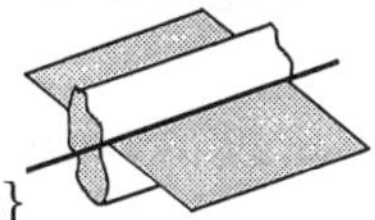 $\{T_D\}$	1	$\{T_D\}$ Unidirectional translation	Any plane parallel to D Any line parallel to D
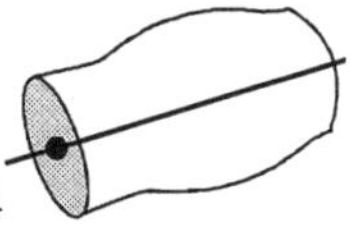 $\{R_D\}$	1	$\{R_D\}$ Rotation around a line with a point	Line D Any point of D
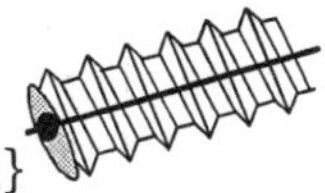 $\{H_{D,p}\}$	1	$\{H_{D,p}\}$ Screw displacement	Line D axis of the screw
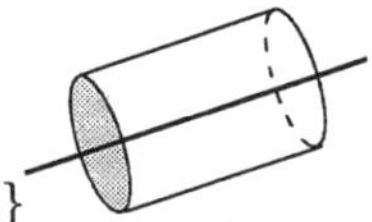 $\{C_D\}$	2	$\{C_D\}$ Actuator displacement	Line D axis of the cylinder
 $\{G_P\}$	3	$\{G_P\}$ Planar displacement	Any plane parallel to P
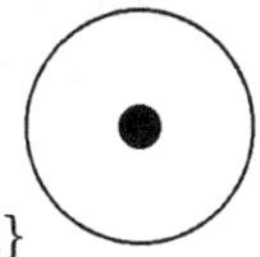 $\{S_O\}$	3	$\{S_O\}$ Spherical displacement	Point O center of the sphere

Table 1 : The 7 classes of surface elements

✧ Class ❶: Invariant surfaces for a spherical displacement.

The MGRS is necessarily the center of the sphere which is the only invariant geometric parameter during a spherical displacement. For such a displacement, all the Cartesian coordinate system having their origin in that point, are none distinguishable, they are said to be equivalent.

✧ Class ❷: Invariant surfaces for a planar displacement.

The MGRS is necessarily a plane which is the only invariant geometric parameter. For such a displacement all the Cartesian coordinate system, using that plane as a coordinate system, are none distinguishable, they are said to be equivalent.

✧ Class ❸: Invariant surfaces for a coaxial displacement of rotation and translation.

The MGRS is necessarily the axis of rotation which is the only invariant geometric parameter. For such a displacement all the Cartesian coordinate system, using that line as a coordinate system, are none distinguishable, they are said to be equivalent.

✧ Class ❹: Invariant surfaces for a helicoidal displacement.

The MGRS is complex; it is made up of a line, the axis of the helix, and this line is oriented by the orientation of the helix. For such a displacement all the Cartesian coordinate system, using that plane and that direction as coordinate system, are none distinguishable, they are said to be equivalent.

✧ Class ❺ : Invariant surfaces for a single-axis rotation.

The MGRS contains a line, (the axis of rotation itself), and a point of this axis. For such a displacement all the Cartesian coordinate system, using that axis as a coordinate system, are none distinguishable, they are said to be equivalent.

✧ Class ❻ : Invariant surfaces for a unidirectional translation.

The MGRS is a line parallel to the direction of the translation, and one of the planes passing through this line. For such a displacement all the Cartesian coordinate system, using that plane and that direction as coordinate system, are none distinguishable, they are said to be equivalent.

✧ Class ❼ : Invariant surfaces for the identity transformation.

The MRGS is a set composed with a plane, a line and a point, defining a complete coordinate system. This system is unique.

As a conclusion, it can be said that a geometric surface is therefore explicitly defined by :

- its geometric type (cylinder, sphere etc.).
- the list of its intrinsic parameters (diameter, form parameters, etc.).
- its class of invariance and its MGRS (positional geometric parameters).

1.3 Classes of surface association :TTRS

If it is now desired to define the relative position of a surface with regards to a surface references, (defined by its geometric type, its intrinsic parameters, and its MGRS), one only has to give the list of its relative positional scalar parameters, which are the relationship between MGRSs. Those relationships between MGRSs will be call CONSTRAINTS.

The concept of **Technologically and Topologically Related Surfaces (T.T.R.S)** had been introduced in order to model those relationships.

A TTRS is either a pair of surfaces or a pair of TTRS or a pair composed with a surface and a TTRS, belonging to the same solid.

Note :

If a surface is set in position with regards to another surface, then it exist 28 combinations which are possible and which represented the 28 association cases of surfaces of the TTRS model.

1.4 Algebraic structure

The surface resulting of the union of two surfaces is a TTRS and is invariant for displacements belonging to a certain displacement subgroup. The class computation of the resulting TTRS is not always an easy operation (it is the intersection of the component subgroups). From that point, all the results are given in extension in table 2. That allows to define reclassifying cases of surfaces: 44 reclassifying cases are identified.

1.5 Oriented relative positioning constraints

Let us now consider all the unions of simple surfaces (whose MGRS is either a point or a line or a plane) that it can be obtained between spherical, cylindrical and plane surfaces. (See table 2, the 6 lower boxes).

The resulting surfaces of those 6 association cases allow the identification of 13 cases of reclassifying.

The MGRSs related to the surfaces used in those 13 reclassifying cases are either points or lines or planes. Those 13 reclassifying cases are defined as being 13 cases of relative positioning between basic surfaces (spherical, cylindrical, plane). Those positioning constraints are listed as follows :

- **C1 : point-point, coincidence,**
- **C2 : point 1 ⇨ point 2, distance,**
- **C3 : point ⇨ plane, distance,**
- **C4 : point-line, coincidence,**
- **C5 : point ⇨ line, distance,**
- **C6 : plane 1 ⇨ plane 2, parallel, distance,**
- **C7 : plane 1 ⇨ plane 2, angle,**
- **C8 : plane-line, perpendicularity,**
- **C9 : plane ⇨ line, parallel, distance,**
- **C10 : plane ⇨ line, angle,**
- **C11 : line-line coincidence,**
- **C12 : line 1 ⇨ line 2, parallel, distance,**
- **C13 : line 1 ⇨ line 2, angle and distance.**

Those 13 constraints constitutes a necessary and sufficient tool allowing to explicitly define the positioning of any surface with regards to any surface using a repeated application of those constraints between the MGRSs of those two surfaces.

9 of the 13 relative positioning constraints are oriented by constructing each time a partial or complete reference Cartesian frame.

2. 13 Oriented Constraints for Dimensioning

It is generally said that it is not possible to completely define the relative positioning of solids using only distance and angle notion.

Using TTRSs, we shall now show that is possible to dimension any complex set of solids, due to the fact, that the TTRS model allows the definition of Cartesian coordinate systems.

First, let us show where sets the problem, in the case of the dimensioning , using the simple example of dimensioning one sphere and two cylinders. The interest of using 13 signed oriented constraints for dimensioning will be shown through that specific example.

Be two cylinders, C_1 and C_2, and a sphere S_1 to position with regards to the cylinders.

In order to perform the dimensioning, the MGRSs of the three entities, i.e. two straight lines D_1 and D_2 and one point M_1 (see figure 1).

	{E}  any surface	$\{T_D\}$ 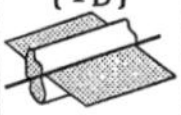 prismatic surface	$\{R_D\}$  revolution surface	$\{H_{D,p}\}$ 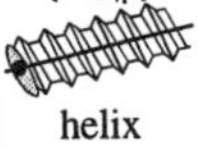 helix surface	$\{C_D\}$ 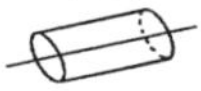 surface cylindrical	$\{G_P\}$ 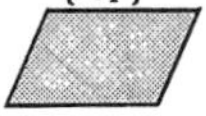 planar surface	$\{S_O\}$ 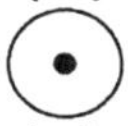 spherical surface
{E} 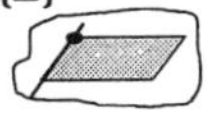	{E}	{E}	{E}	{E}	{E}	{E}	{E}
$\{T_D\}$ 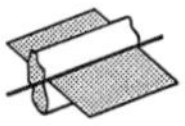		$D_1 // D_2$ ➡ $\{T_{D1}\}$, otherwise ➡ {E}.	{E}	{E}	$D_1 // D_2$ ➡ $\{T_{D1}\}$, otherwise ➡ {E}.	$D_2 // P_1$ ➡ $\{T_{D2}\}$, otherwise ➡ {E}.	{E}
$\{R_D\}$ 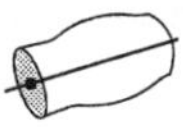			$D_1 = D_2$ ➡ $\{R_{D1}\}$, otherwise ➡ {E}.	{E}	$D_1 = D_2$ ➡ $\{R_{D1}\}$, otherwise ➡ {E}.	$D_2 \perp P_1$ ➡ $\{R_{D2}\}$, otherwise ➡ {E}.	$O_1 \in D_2$ ➡ $\{R_{D2}\}$, otherwise ➡ {E}.
$\{H_{D,p}\}$ 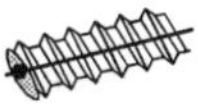				$\left\{\begin{matrix} D_1 = D_2 \\ p_1 = p_2 \end{matrix}\right\}$ ➡ $\{H_{D1,p1}\}$, otherwise ➡ {E}.	$D_1 = D_2$ ➡ $\{H_{D1,p1}\}$, otherwise ➡ {E}.	{E}	{E}
$\{C_D\}$ 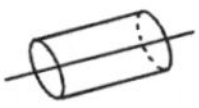					**C11 :** $D_1 = D_2$ ➡ $\{C_{D1}\}$, **C12 :** $\{D_1 // D_2$, [illegible]$\}$ ➡ $\{T_{D1}\}$, **C13 :** otherwise ➡ {E}.	**C8 :** $D_2 \perp P_1$ ➡ $\{R_{D2}\}$, **C9 :** $D_2 // P_1$ ➡ $\{T_{D2}\}$, **C10 :** otherwise ➡ {E}.	**C4 :** $O_1 \in D_2$ ➡ $\{R_{D2}\}$, **C5 :** otherwise ➡ {E}.
$\{G_P\}$ 						**C6 :** $P_1 // P_2$ ➡ $\{G_{P1}\}$, **C7 :** otherwise ➡ $\{T_D\}$.	**C3 :** $\{R_D\}$
$\{S_O\}$ 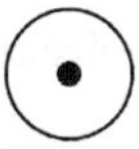							**C1 :** $O_1 = O_2$ ➡ $\{S_{O1}\}$, **C2 :** otherwise ➡ $\{R_D\}$.

Table 2 : Internal commutative composition law

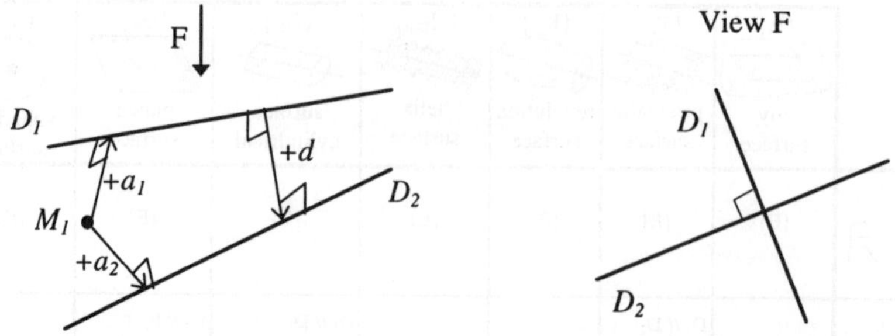

Figure 1 :Two cylinders and one sphere represented using their respective MGRSs

That figure also shows the three relative positioning constraints which it may be established between those three elements :

- the oriented constraint between D_1 and D_2, C13 : Line D_1 ⇨ Line D_2 angle and distance. That constraint is associated with one distance signed parameter, + d, and with one angular parameter equal to *90°*.
- the oriented constraint between M_1 and D_1, C5 : Point M_1 ⇨ Line D_1 angle and distance. That constraint is associated with one distance signed parameter, + a_1.
- the oriented constraint between M_1 and D_2, C5 : Point M_1 ⇨ Line D_2 angle and distance. That constraint is associated with one distance signed parameter, + a_2.

The point *M1* position is not completely determined since, in order to satisfy the two constraints in which that point is implied, that point has to be on the intersection curve of the two cylinders.

The point *M1* keeps **one degree of freedom**, thus it is not completely located. Without the use of the 13 signed relative positioning constraints of the TTRSs model, it is not possible to clearly express that the set made of the two cylinders constitutes **one solid**.

In order that the positioning be completed, the **resulting MGRS** of the TTRS ($C_1 \cup C_2$) has to be constructed.

Note that either for distance values, which equal to 0, or for angles equal to 90°, then the parameter is not signed, therefore the constraint is not oriented.

❶ The line Δ is defined as being perpendicular to both lines D1 and D2, using the declaration of the two following constraints :

- C13 : Line *Δ* ⇨ line D_1, angle and distance, such that the distance equals to 0 and that the angle equals to 90°;
- C13 : Line *Δ* ⇨ line D_2, angle and distance, such that the distance equals to 0 and that the angle equals to 90°.

❷ The plane P_1 is defined as containing both lines D_1 and Δ, and using **the declaration** of the two following constraints :

- C9 : Line D_1 ⇨ plane P_1, parallel, distance, such that the distance equals to 0;
- C9 : Line Δ ⇨ plane P1, parallel, distance, such that the distance equals to 0.

❸ The point O1 is defined as being the intersection point of the lines D1 and Δ, using the declaration of the two following constraints :

- C4 : Point O_1 ⇨ line D_1, coincidence;
- C4 : Point O_1 ⇨ line Δ, coincidence.

In that case the position specification of the point M_1 can be completed by the declaration of an oriented relative positioning constraint using one those three constructed elements :either O_1 or Δ or P_1.

For instance it can be chosen to declare, as follows :

- C3 : Point *M1* ⇨ plane P_1, distance, such that the distance equals to $+b1$.

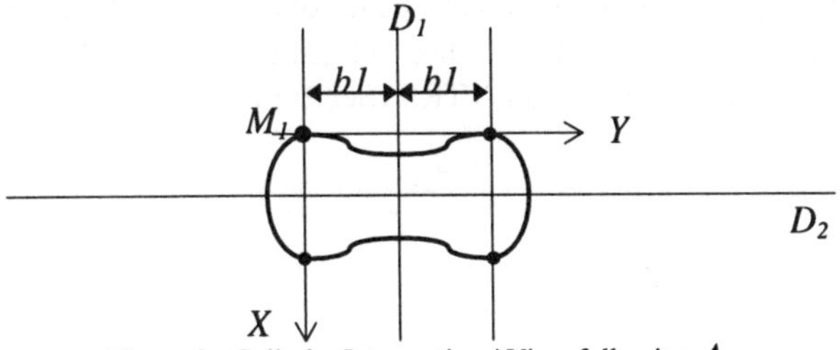

Figure 2 : Cylinder Intersection / View following Δ

Figure 2 shows the existence of **four possibilities** regarding to the point M_1 position.

It is to notice that the choice of one of the four points will have for repercussions the orientation of the resulting MGRS of the considered TTRS. That orientation will be following the two directions $\overrightarrow{X}$ and $\overrightarrow{Y}$, so as to be in conformity with the signs given to the positioning constraints (see figure 2).

Thus, the TTRS model for the dimensioning of geometrical features is essentially constituted with the declaration of two relationship types :

- the association relationship between the surfaces, which themselves define the TTRS;
- the relationships corresponding to the oriented constraints of positioning between the geometrical elements of situation, which define the TTRS dimensioning.

Use such a dimensioning model implies the handling of parameters intrinsic to the geometrical objects and which are thus not depending of a possible Cartesian datum of work.

Therefore, the TTRS model for the dimensioning is completely parametrics.

3. 13 Oriented Constraints for Tolerancing

3.1 Tolerancing model

In accordance with the ISO 10303-47 (FDIS stage), part 47 of STandard for Exchange of Product model data, the definition of a geometrical tolerance is based on the relative positioning of the toleranced element and of the tolerance zone with regards to the coordinate system.

Any tolerance zone is positioned with regards to the toleranced element using one or several constraints of the list of the 13 signed constraints. That set (tolerance zone $\cup$ toleranced element) constitutes the toleranced TTRS, which is itself positioned with regards the TTRS datum using one or several constraints of the list of the 13 signed constraints. That other set (toleranced TTRS $\cup$ TTRS datum) constitutes the geometrical tolerance TTRS (See figure 3).

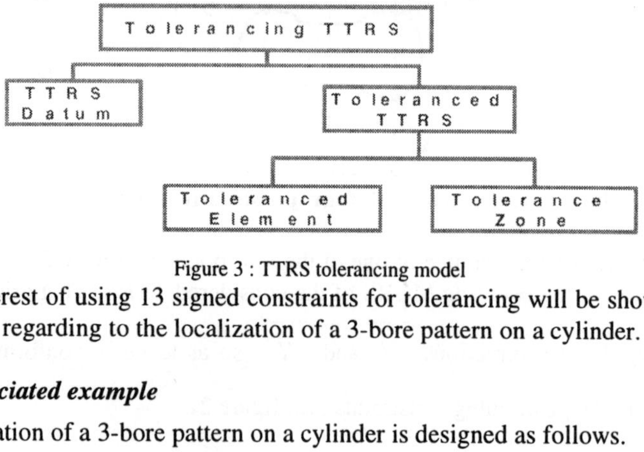

Figure 3 : TTRS tolerancing model

The interest of using 13 signed constraints for tolerancing will be shown through an example regarding to the localization of a 3-bore pattern on a cylinder.

3.2 Instanciated example

The localization of a 3-bore pattern on a cylinder is designed as follows.

The tolerance zone construction increases following three main stages :

- the datum or coordinate system definition,
- the tolerance zone construction and
- the tolerance zone positioning with regards to the datum.

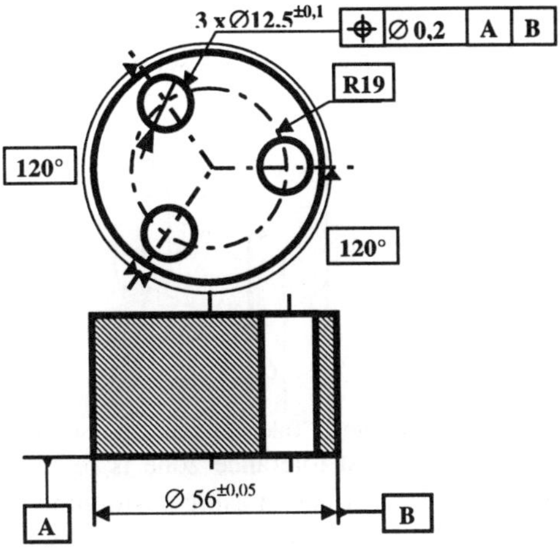

Figure 4 : Bore localization

3.2.1 *Datum definition*

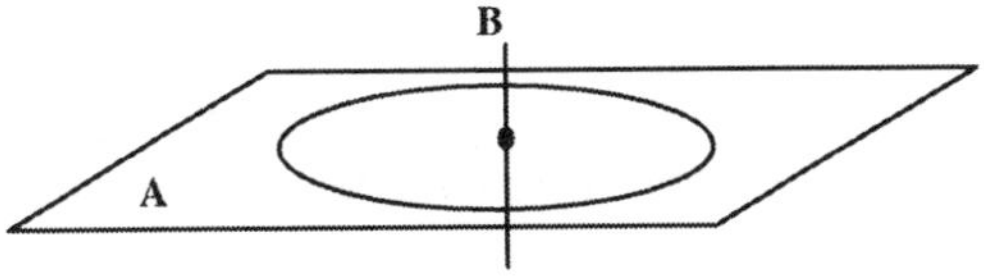

Figure 5 : Datum definition

The Euclidean datum is the union of the plane A and of the cylinder B. The TTRS datum is the association of the MGRS plane A and of the MGRS line B. The plane A is the primary datum and the cylinder B is the secondary datum.

The MGRS line B is positioned with regards to the MGRS plane B under the positioning constraints C8, plane A-line B perpendicularity. By that way, the perpendicularity implicitly expressed in the design draft, is now **explicitly** expressed.

However, such a constructed datum is not complete. It means that the datum, made of the union of the plane A and of the line B, keeps two degrees of freedom. Here these degrees of freedom are a rotation around the line B and a translation along that same line B.

Thus derived MGRS are constructed. The intersection point, the between the plane A and the line B, is constructed. The derived MGRS point is constructed using the positioning constraint C4, point (to construct)-line B coincidence, and the positioning constraint C3, point (to construct) ⇨ plane A distance.

Then, the tolerance zone construction can be realized and that is done with regards to the datum.

3.2.2 *Tolerance zone construction*

In that case of study, the tolerance zone is a set of three tolerance zones, as shows on figure 6.

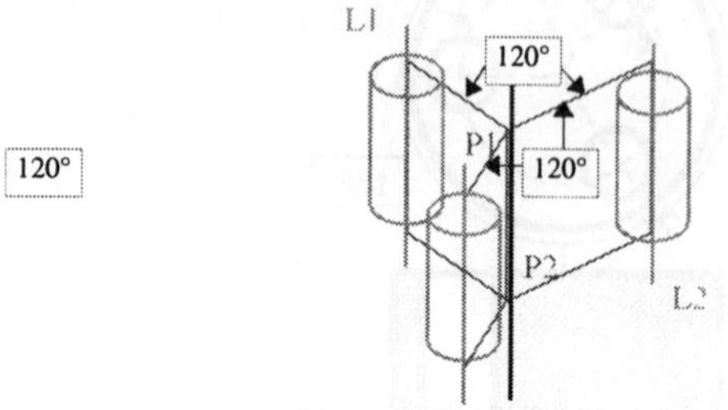

Figure 6 : Tolerance zone construction

The construction of the first tolerance zone is realized in two stages. The first stage consists in the axis 1 construction using positioning constraints :

- C12 : axis ⇨ line L1, parallel, distance and
- C9 : line L1⇨ plane P1, parallel, distance, which here equals to 0.

The second stage is the zone construction using C5, any point ⇨ line L1 distance.

That tolerance zone construction constitutes one scenario. That scenario is reused in order to construct the second and the third tolerance zone.

The positioning constraint C7, plane P1 ⇨ plane P2, angle such that the angle equals 120°, is applied to construct the plane then the scenario 1 and to obtain the complete second tolerance zone.

The same procedure is to follow in order to obtain the third tolerance zone. Those three tolerance zones together constitute the final tolerance zone of the schema.

3.2.3 *Tolerance zone positioning with regards to the datum*

The tolerance zone is positioned with regards to the datum under the constraint C8, plane A ⇨ line L1 perpendicularity and C7, plane P1 ⇨ plane A, angle such that the angle equals 90°, (see figure 7).

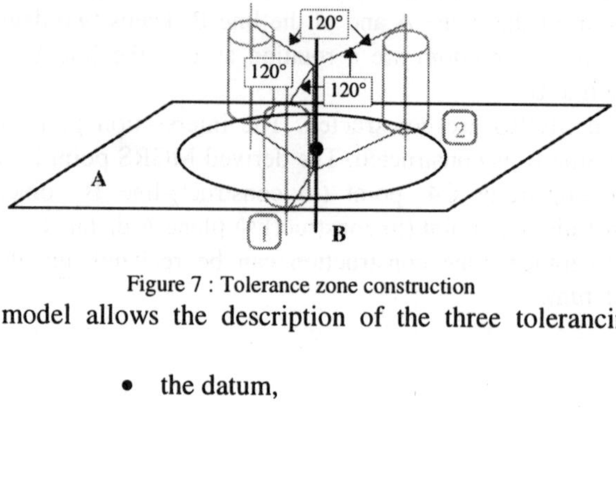

Figure 7 : Tolerance zone construction

The TTRS model allows the description of the three tolerancing component groups :

- the datum,

- the toleranced element and
- the tolerance zone.

But that tolerancing model also explicitly describes the relative positioning of those component groups the one to each over. By that way, the nominal 3D-geometry is explicitly described.

It is main to precise that at this stage of study, the TTRS tolerancing model furnishes an explicitly described softgage. From that point of view, main changes are realized regarding to the inspection : the inspection is no more developed with regards to the machined part drawing (see figure 4), which has to be interpreted by the inspector, but with regards to a data structure, which is from a semantic point of view far too rich.

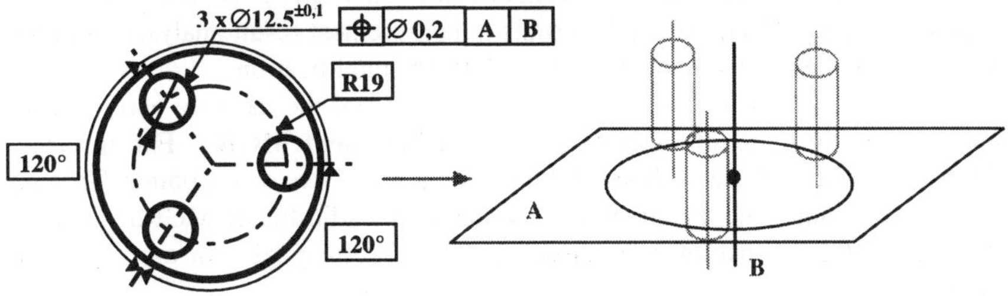

Figure 8 : Machine part draft with regards to softgage

Consequently having that softgage, it is not necessary any more to proceed in successive construction stages, of nominal entities (such as the datum, the toleranced element and the tolerance zone) using relative positioning constraints, since the information already explicitly exists in the data structure.

That concept will not only suppress or limit the interpretation problem on the subject, regarding to the tolerance zone construction, but also the construction errors of the datum.

The global quality of the dimensional inspection seems greatly improved. In the next section, the new TTRS concept applied to the metrology field, and more precisely to the dimensional inspection, is treated in details.

4. 13 Oriented Constraints for Metrology

4.1 Input data for dimensional inspection

The model structure allows to **automatically** perform dimensional inspection constructions usually made by the inspector. The interest of using 13 signed constraints for metrology is shown at the next section through the example regarding to the localization of a 3-bore pattern on a cylinder, seen on figure 8.

The inspection is composed with two distinct and successive stages : the **probing** and the **measure**.

The goal is to use the dimensioning information and the tolerancing information as input data for the probing activity. During the course of the probing activity, probed point data are supplied. The global aim of that activity is, of course, to associate a probed point data with each nominal surface, which are necessary to inspect a given tolerance.

That probed point data constitutes itself an input data for the measure activity. However those point data do not form a **sufficient** information in order to correctly proceed in the measure activity. Both information relative to dimensioning and tolerancing are **necessary**. Following this idea, the global TTRS tolerancing model has to be use as an input data for the correct course of the measure activity.

Afterwards, that measure activity supplies measurement results. Then the information feedback not only to the manufacturing engineering but also back the design engineering. That last back permits a measurement result analysis, in order to lead to possible production adjustments and design modifications.

However, the progress of the sensor activity and of the measure activity can be alternative in the case of determination of **adjustment MGRS**. For instance, a cylindrical surface can be, at first, probed using a set of five or six points. That set is fitted to determine one cylinder axis which is an adjustment MGRS. Then that adjustment MGRS constitutes the direction of the probing axis, for a probing along an axis.

That determination case of adjustment MGRS is also encountered for the irregular surfaces, by the determination and the construction of local coordinate system or datum

4.2 Dimensional Inspection Proposal

In this proposal, the dimensional inspection is composed of two stages, the probing activity and the measure. That measure phase is itself composed with three successive stages :

- the fitting of the actual datum,
- the put into coincidence of both the actual datum and the nominal datum and
- the measurement and/or the control.

4.2.1 Actual datum fitting

The actual datum fitting is no more released using successive construction stage, but using the relative positioning constraints.

Here, the example refers to figure 8. The following elements are fitted :

➊ the datum plane A', using the probed point data relative to the surface A, (see figure 9),

❷ the datum cylinder B', using the probed point data relative to the surface B. That **fitting** is realized **under the constraints** of positioning C8, line B ⇨ plane A perpendicularity.

Note that the TTRS tolerancing model allows to clearly clarify :

❶ the **hierarchy** between the datum components. In that case of study ; the surface A' is the primary datum and the surface B' is the secondary datum.

❷ the **perpendicularity constraint** between the datum plane A' and the datum axis cylinder B'.

This method avoids the interpretation problems that the inspector may encountered since the information is already clarified.

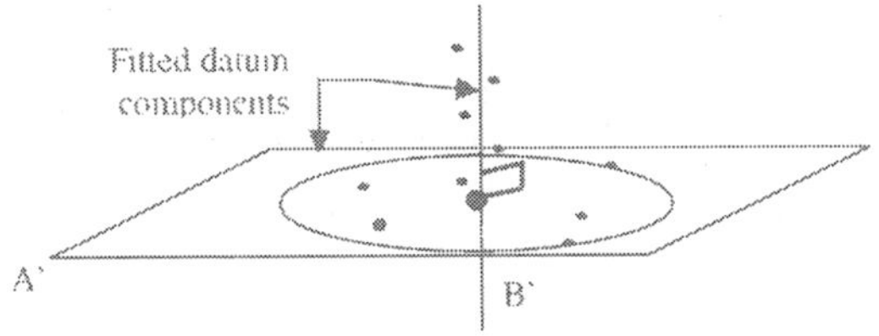

Figure 9 : Actual datum fitting

4.2.2 Actual and nominal datum coincidence

At that stage of study, the model provides on one hand, the nominal datum or the softgage and on the other hand, the fitted actual datum, as figure 10 shows. The method consists in the put into coincidence of the two systems, as figure 11 shows.

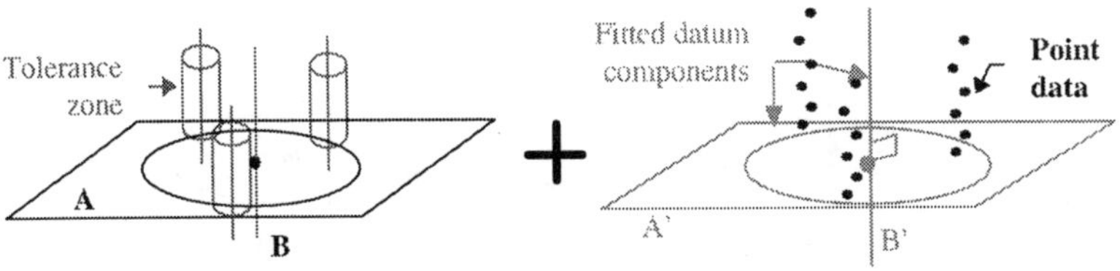

Figure 10: The nominal datum or softgage and the actual datum

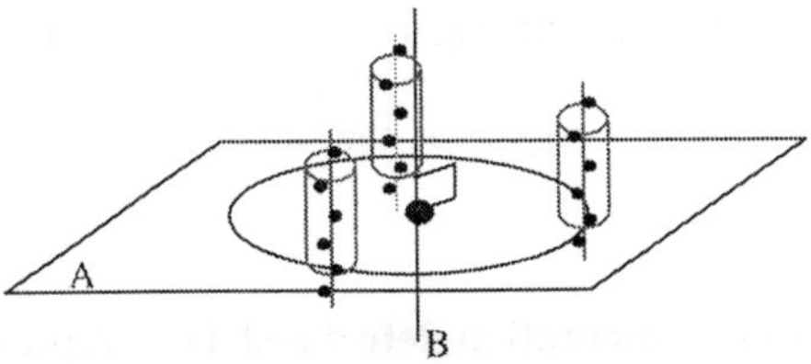

Figure 11: Nominal and actual datum coincidence

In that example, there is still a degree of freedom, which is the rotation around the line B. But the point data to measure and to control are clearly positioned. The measurement and the control can be realized.

4.2.3 Measure and Control

If a **control** is desired, it has to be verified if the point data go or not in a gage. If a **measure** is desired, the distances d_i, shown at figure 12, have to be smaller or equal to the half of the tolerance noted on the machined part definition drawing. The measure of those distances can be expressed using the positioning constraint C5, point M_i ⇨ line, distance.

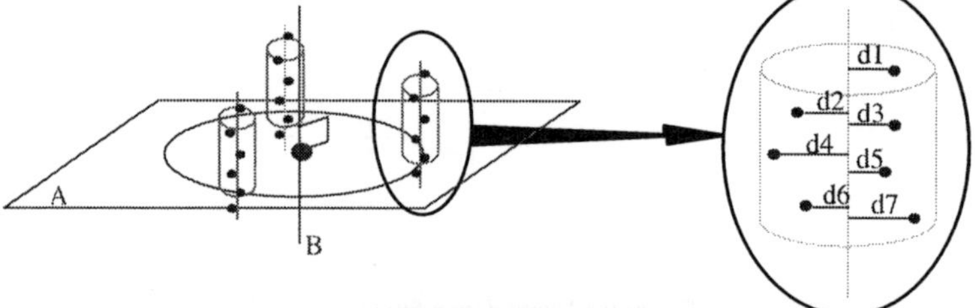

Figure 12: Measurement

In that instance, that measure and that control allow to inspect the functionality of a machined part. However the TTRS tolerancing model allows also to detail in-process inspection. In the previous example, consider separately each bore. Then the actual axis of the chosen bore is fitted.

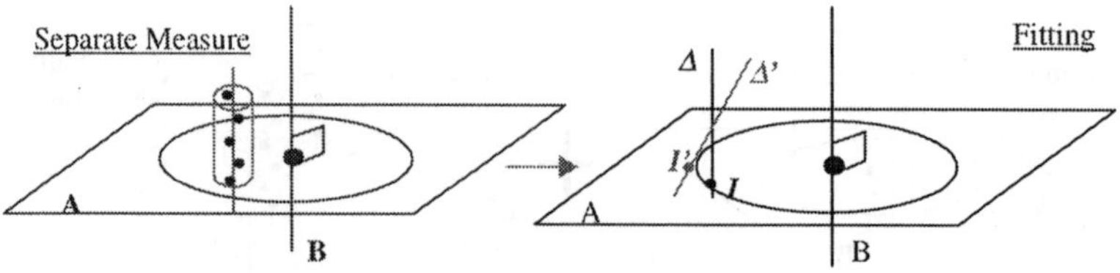

Figure 13: Separate measure

The points I and I' belong to the datum surface. The vectorial gap of the line Δ, belonging to the nominal datum, with regards to the line Δ', belonging to the actual datum can be expressed using a torsor of the small displacements : $\left\{ \begin{matrix} \overrightarrow{D_{\Delta\Delta'}} \\ \overrightarrow{\Omega_{\Delta\Delta'}} \end{matrix} \right\}$.

$\Omega_{\Delta\Delta'}$ precisely represents the **orientation defect** and $\overrightarrow{D_{\Delta\Delta'}}$ represents the **translation defect**.

That information may imply a correction of the tool regarding to the translation and/or the rotation. The goal is to better transmit the information to the producer regarding to the production defects. The mean use of that information is the specification of the vectorial gap of each one of the three bores with regards to the nominal datum.

Regarding to the irregular surfaces, a fitting is realized under the constraint of the minimum shape defect.

The interest is put on the vectorial gap existing between the nominal MGRS, related to the nominal irregular surface and the MGRS, related to the actual irregular surface. By the same way, the vectorial gap of the surface actually obtained, with regards to the nominal aimed surface, is expressed using a tensor of the small displacements. Those metrological results should facilitate the exchanges between the metrologists and the producers and afterwards the adjustment of production tools.

5. Conclusion

The TTRS model allows the explicit dimensioning and positioning of any surface with regards to any other surface using constraints of the list of the 13 constraints and permits the discrimination of the solutions using the orientation of the constraints.

In the tolerancing field, the TTRS model also explicitly describes and locates, with regards to its own datum and using signed constraints, the tolerance zone of any element. In order to inspect those tolerances, the TTRS model finally permits to automatically perform dimensional inspection constructions.

That is at one disposal an integrated model which allows completion, consistence and explicit definition in the dimensioning, tolerancing and metrology fields.

References

1. A. CLEMENT *et al*, 3-D Tolerancing with CAD-CAM systems, Addison Wesley, (1996).
2. A. CLEMENT. Resolution of Positioning Solids. Annals of the CIRP, vol. 40/1/1991.
3. P. DUFOSSE. Automatic Dimensioning and Tolerancing. 3rd CIRP Seminar on Computer Aided Tolerancing, April 1993.
4. **ISO/DIS 10303-47:**« Industrial automation systems - Product data representation and exchange - Part 47: Integrated generic resources: Shape variation tolerances », ISO, 1995.
5. A. BALLU, L. MATHIEU. Analyses of dimensional and geometrical specifications: standards and models. 3rd CIRP Seminar on Computer Aided Tolerancing, April 1993.

6. B. CHARLES, A. CLEMENT, A. DESROCHERS, P. PELISSOU, A RIVIERE. Toward a computer aided functional tolerancing model. International Conference on CAD/CAM and AMT in Israel, CIRP. Sessions on tolerancing for function in a CAD/CAM environment, Proceedings, vol.2, Jerusalem, Israel, 11-14 December 1989.
7. K.W. CHASE, W.H. GREENWOOD. Design issues in mechanical tolerance analysis. Advanced Topics in Manufacturing Technology: Product Design, Bioengineering and Space Communication, ASME Winter Annual Meeting, Boston, MA, December 13-18, 1987.
8. A. CLEMENT, A. DESROCHERS, A RIVIERE. Theory and practice of 3D tolerancing for assembly. 2nd CIRP International Working Seminar on Computer Aided Tolerancing, May 1991.
9. H.A. ELMARAGHY, W.H. ELMARAGHY. A system for modeling geometric tolerances for mechanical design. 3rd CIRP Seminar on Computer Aided Tolerancing, April 1993.
10. A. WIRTZ. Vectorial Tolerancing a basic element for quality control. 3rd CIRP Seminar on Computer Aided Tolerancing, April 1993.
11. F. ETESAMI. Tolerance verification through manufactured part modeling. Journal of manufacturing systems, vol. 7, N. 3, 1988.
12. P. BOURDET, A. CLEMENT. Controlling a Complex Surface with a 3 Axis Measuring Machine. Annals of the CIRP vol. 25/1/1976, pages 354 - 361.
13. B. CHARLES, A. CLEMENT, A. DESROCHERS, P. PELISSOU, A RIVIERE. Controlling a mechanical part designed with a feature based system. Computer Integrated Quality System in CIM, IFIMT.C.S./W.G.5.3. Conference, Belgrade, Yugoslavia, 20-23 June 1989.

Advanced Mathematical Tools in Metrology III
Edited by P. Ciarlini, M.G. Cox, F. Pavese & D. Richter

GRADED REFERENCE DATA SETS AND PERFORMANCE PROFILES FOR TESTING SOFTWARE USED IN METROLOGY

M.G.COX
Centre for Mechanical and Optical Technology, National Physical Laboratory Teddington, Middlesex TW11 0LW, UK

The concern here is with the use of reference data sets as the basis of black-box testing of software used in metrology for measurement data processing. Graded data sets, i.e., sequences of reference data sets where the members in a sequence depend on a set of increasing parameter values representing problem "degree of difficulty", are discussed. It is shown how these data sets can be generated in such a way that the corresponding reference solutions can be specified *a priori*. Consideration is given to the use of these data sets to produce performance profiles of test software, the use of which can help in identifying deficiencies in the software.

1 Introduction

Software used in metrology for processing measurement data can fail because it implements algorithms that are unreliable or not robust. Algorithms can be fragile for a number of reasons, such as (i) failure to scale, normalise or otherwise transform the input data appropriately before solution (and to perform the inverse operation following solution), (ii) the use of an unstable parametrisation of the problem, (iii) the use of a solution process that exacerbates the inherent (natural) ill-conditioning of the problem, and (iv) a poor choice of formula from a set of mathematically (but not numerically) equivalent forms.

Instances of all types of fragility prevail but, unless the source code is available or the software supplier provides sufficiently detailed documentation, it is necessary to rely entirely on black-box testing to discern whether any particular item of software is fit for purpose. The concern here is with reference data sets for the black-box testing of software used in metrology. Although such software is very widely used the state of the art of metrology software testing is somewhat immature. However, useful progress is being made on the development of approaches for testing the individual software modules used directly by metrologists or as part of software packages used in metrology.[3] (It is felt that until a widely recognised approach to testing individual modules is available, the testing of integrated systems remains a very difficult task.)

Black-box testing involves submitting reference data sets to software under test, the *test software*, and the resulting output, the *test results*, is compared with known reference results. The emphasis here, however, is on techniques for *generating* pairs of reference data sets and corresponding reference results, and on demonstrating the *use* of this basic concept and *extensions* of it.

Section 2 of this paper covers some of the main considerations in designing reference data sets and Section 3 the use of these data sets. Section 4 is concerned with the production of reference data sets. Section 5 presents examples of application. Section 6 contains our conclusions and possible future activity.

2 Design of reference data sets

Some of the main considerations in designing reference data sets for checking the fitness for purpose of software are the identification of possible deficiencies, the imposition of known properties in the data sets, prior knowledge of the solution, and the need to mimic actual measurement data sets.

Checking fitness for purpose can partly be accomplished by ensuring that the software operates correctly for data sets that are *representative* of the application. In addition, it is important to include data sets designed to reveal the extent to which the software can yield sufficiently accurate solutions. For both these purposes it is highly desirable that the data sets possess certain properties. One valuable property is that the data sets possess *known* solutions.

Software may be of *general usage*, e.g., for determining a minimum of a general function of any number of variables. Such software appears in mathematical software libraries and, because the extent of its use cannot be anticipated, the degree of testing should be considerable. Such testing is beyond the scope of this paper. The software may be of more *specific* use, e.g., to fit Gaussian associated circles to measured data points.[8] The methods discussed here are applicable to such software.

Software may be deficient in diverse ways, such as an inability to deal accurately with measurements containing a significant noise component, large numbers of measurement points, or certain distributions of points. Suitable tests can be devised to help reveal such deficiencies.

Real measurement data sets have certain properties that in total can be quite difficult to characterise. However, data sets can be devised which are "close" to such data sets *and* which have known solutions.[5]

3 Use of reference data sets

This section treats the factors influencing the comparison of reference results and test results, and indicates the quality metrics used for this purpose.

Some of the major factors affecting the comparison of test and reference results are the computational precision η of the arithmetic used to produce the test results, the numerical precision of the reference data sets, the degree of difficulty K and scale of the problem, and the problem parametrisation. We consider the effects of η and K, assuming that problem scale effects have been incorporated into K, and that the reference results are known to adequate accuracy for comparison purposes. Further information concerning these factors and their influence and other effects such as input-output is available.[3,4]

Following the use of each reference data set three quality metrics which indicate how the test software performed on that data set can be determined[3]: (i) the difference between the test and reference results, an *absolute* measure of departure, (ii) the number of figures of agreement, a *relative* measure of departure, and (iii) a *performance measure*. We concentrate here on the use of (iii). It is indicated below how this metric is determined and used in practice.

Let $\mathbf{d} = \mathbf{b}^{(\text{test})} - \mathbf{b}^{(\text{ref})}$, where $\mathbf{b}^{(\text{test})}$ denotes the test results and $\mathbf{b}^{(\text{ref})}$ the reference results. We use $d(\mathbf{b}) = \text{RMS}(\mathbf{d})$ to measure this difference, where $\text{RMS}(\mathbf{x}) = \|\mathbf{x}\|_2/n$ denotes the root-mean-square value of an n-vector $\mathbf{x}$.

For each data set the problem degree of difficulty K is determined such that $d(\mathbf{b})$ can sensibly be compared with a tolerance equal to $K\eta$. Thus, K depends on the problem condition and scale. The quantity $d(\mathbf{b})/(K\eta)$ will then have the property that it is of order unity for software that performs near-optimally and is possibly very large for poorly-performing software.

The performance measure $P(\mathbf{b})$ is defined by

$$P(\mathbf{b}) = \log_{10}\left(1 + \frac{d(\mathbf{b})}{K\eta}\right). \tag{1}$$

It indicates the number of figures of accuracy *lost* by the test software over and above what software based on an optimally stable algorithm would produce, being near zero if $\mathbf{b}^{(\text{test})}$ and $\mathbf{b}^{(\text{ref})}$ agree as closely as could be expected, and having a value of about k if the agreement is k figures less than this. A related performance measure is used in testing software for evaluating special functions.[15]

4 Production of reference data sets

This section is concerned with the use of data generators to produce reference data sets, leading to the *null-space* approach to data generation (based on solution characterisation). It also covers graded data sets and performance profiles.

A *reference pair*, i.e., a reference data set and the corresponding reference results, may be produced in two ways: (i) start with a reference data set and apply reference software[2] to it to produce the corresponding reference results and, (ii) start with some reference results and apply a *data generator* to them to produce a corresponding reference data set. Note that (for problems with a unique solution) there is one set of reference results corresponding to a given reference data set. Conversely, for given reference results there is in general an infinite number of corresponding reference data sets. As a simple example, there is a unique standard deviation σ for a given set of two or more numbers, whereas given σ there are infinitely many data sets having this value as their standard deviation. We return to this example below and in Section 5.2.

Null-space methods[4,5] provide a facility for generating *families* of data sets such that, when any of these sets is submitted to correctly working software, nominally the same solution is produced which, moreover, is identical to a known solution. A further advantage of null-space methods is that they can be used to characterise the *space of reference data sets* having the given solution. In particular, a sequence of data sets from this space can be extracted having a range of degrees of difficulty that are useful for software testing.

We illustrate the null-space approach for the linear least squares model $\mathbf{y} = A\mathbf{b}+\mathbf{e}$, where A denotes the observation matrix, $\mathbf{y}$ the observation vector, $\mathbf{b}$ the vector of model parameters and $\mathbf{e}$ the vector of residuals. Now the least squares solution is characterised by $A^{\mathrm{T}}\mathbf{e} = 0$.[11] Thus, if N is a basis for the null space of A^{T}, i.e., $A^{\mathrm{T}}N = 0$, then for a vector $\mathbf{r} = N\mathbf{u}$ for *any* choice of vector $\mathbf{u}$, the replacement of $\mathbf{y}$ by $\mathbf{y} + \mathbf{r}$ will leave the solution vector $\mathbf{b}$ unchanged. This result follows from pre-multiplication of $\mathbf{y} + \mathbf{r} = A\mathbf{b} + \mathbf{e} + \mathbf{r}$ by A^{T} and using the above result. Thus, from any one data set any number of further data sets having the same solution can readily be constructed.

Consider what is arguably the simplest possible example, the mean and standard deviation of a set of numbers. This example is closely related to the above model, since determining the mean of a set of numbers is equivalent to fitting them by a constant, and their standard deviation is identical to the root-mean-square residual of the fit. Any two values having $\bar{x}$ as their mean are given by $\bar{x} \pm \lambda$ for any λ. (This is the only null-space perturbation possible in this case.) Further, any two values given by $\mu \pm s/\sqrt{2}$ have s as their standard deviation for any value μ. Thus, by varying μ, for example, different data sets of size two can be generated where in each case the solution standard deviation is s.

A well-known spreadsheet package loses 12 of the 14 decimal figures available when $s = 0.1$ and $\mu = 10^6$, whereas another loses just three figures. When μ is increased to 10^7, the first spreadsheet returns *zero* (!) as the answer (as a result of losing *all* figures), whereas the second delivers a value differing from the true value by less than 10^{-9}. Thus, μ (or λ) can be used as a *performance parameter*, i.e., a varying parameter to investigate performance.

These data sets contain only two values, and in each case there are just *two* solution parameters, $\bar{x}$ and s. Null-space methods [5] permit the above concept to be extended to any number of values. Moreover, they permit not just simple problems such as that above to be handled, but extend to any number of solution parameters and to very complicated problems, even involving nonlinear optimisation with constraints.

Much software used in metrology has the (unfortunate) property that it works well for certain data sets but produces unacceptably inaccurate results for other sets. An example is polynomial regression software, where the results of fitting a polynomial of low degree can be perfectly acceptable whereas those for higher degrees would be contaminated by errors introduced by the solution algorithm. This example is considered in more detail in Section 5.3.

The testing of software on a small number of data sets can provide valuable information about the software, particularly if the sets are selected to be as disparate as possible in terms of their locations in the space of possible data inputs. Such data sets provide a number of "spot checks" of the test software, and are very useful in the sense that the software can be regarded as deficient if it fails any one of them. Data sets of this form have been used for a number of years in testing software used in dimensional metrology for fitting associated features such as spheres, cylinders and cones to coordinate measuring machine data [6] and will become more formally established when an ISO standard is published [13], and in analytical chemistry. [3,7]

In order to quantify the behaviour of metrology software in such cases, use can be made of the concept of *graded* data sets. The idea is to have a *sequence* of data sets, where each data set in the sequence represents a problem that in some sense is more difficult than the previous member in the sequence. A single suite of graded data sets explores *one dimension* of the problem addressed by the software, but the resulting *performance profile*, the *graph* of the values of P against the degree of difficulty K (or some related performance parameter), can provide valuable information about the software. If, for *graded reference data sets* (taken in order), the values of the resulting performance measure $P(\mathbf{b})$ have a tendency to *increase*, it is likely that an unstable method has been employed. (Performance profiles in a somewhat different setting have also been studied by Lyness. [14]) The performance profile will generally be expressed as a series of points (or the straight-line segments joining them). By introducing variability into each data set for each value of the performance parameter, e.g., by using random numbers in a controlled way to simulate measurement error, the performance profile will become a *swathe* of points, which can be advantageous in the fair comparison of rival algorithms.

5 Examples of application

We give four examples of application, which have arisen in analytical chemistry and dimensional metrology, although all examples have broader application.

5.1 *The sample geometric mean*

The *geometric mean* of m numbers $x_i > 0$, $i = 1, \ldots, m$, is defined by

$$\bar{x}_g = \left(\prod_{i=1}^{m} x_i \right)^{1/m}. \tag{2}$$

Most software for computing $\bar{x}_g$ will produce values for modest values of m that have a relative accuracy approaching η. However, for large m the *same* software can fail because of overflow or produce *zero* due to underflow.

Consider a roundness measuring instrument which provides suppressed radius values at say 1000 points on a section of a nominally cylindrical workpiece. It is required to determine the geometric mean $\bar{x}_g$ of these values. If in this application each x_i is approximately equal to 2 μm the direct formation of the product in (2) would yield a value of the order of 10^{300} which would cause overflow on many computers, although $\bar{x}_g$ is of course of order 2 μm. If instead each x_i were approximately equal to 0.5 μm, formation of the product would *underflow* and the value *zero* rather than about 0.5 μm would be returned.

Since the geometric mean is identical to the arithmetic mean in a transformed variable, i.e.,

$$\ln \bar{x}_g = \frac{1}{m} \sum_{i=1}^{m} \ln x_i, \tag{3}$$

the use of (3) is a viable alternative approach: overflow and underflow become virtually impossible and a value having relative accuracy approaching η is returned (although the use of the ln and exp functions can introduce some small additional loss of accuracy).

It is straightforward to design sequences of data sets for testing geometric mean software for the above frailties. Define data $X_k = \{10k$ points equispaced in the interval $c \pm 100k\eta\}$. Software known to implement (2) directly and software based on (3) were applied to data sets X_k, $k = 1, \ldots, 40$, in the cases $c = 1.99$ and $c = 0.499$ (i.e., values near two and one-half). Performance measures $P(X_k)$ based on (1) for computing $\bar{x}_g = \bar{x}_g(X_k)$ are shown in Figure 1. The performance was consistent with expectation. The software based on (3) produced $P(X_k) \approx 0$ for both values of c and for all k (indicated by plus signs in Figure 1). For $c = 1.99$, software based on the direct use of

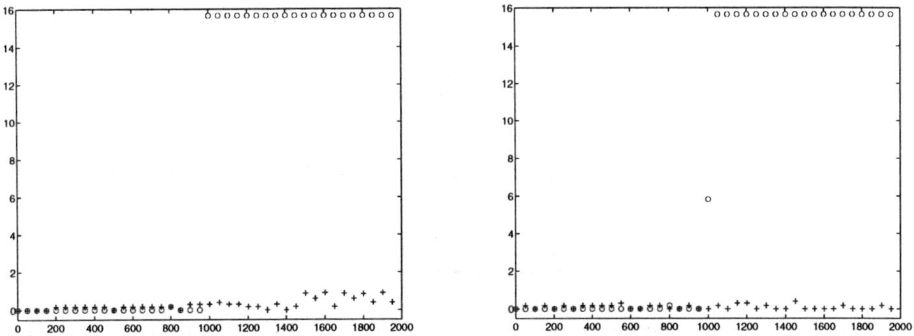

Figure 1: Performance profiles for the geometric mean computed using a *robust* algorithm (indicated by plus signs) and a *fragile* algorithm (small circles). The horizontal scale is proportional to the number of points contributing to the mean and the vertical scale is the performance measure. The left panel show the profile for the "overflow data" and the right panel that for the "underflow data"

(2) (indicated by small circles) produced $P(X_k) \approx 0$ for all $k \leq k'$ and *failed* for all $k > k'$, where k' depends on the overflow threshold of the computer. Similarly, for $c = 0.499$, this software produced $P(X_k) \approx 0$ for all $k \leq k'$ (possibly a different k' from that above) and $P(X_k) \approx -\log_{10} \eta$ (loss of all significant figures) for all $k > k''$ ($> k'$). For $k' < k \leq k''$ the effects of *gradual underflow* as implemented in the IEEE arithmetic[12] used are apparent in the right panel of Figure 1 for the P-value for the fragile method corresponding to approximately 1000 points. This value of approximately 6 as opposed to the values of approximately zero and 16 is a result of the decreasing loss of accuracy. (More such values are observed if the performance profile is plotted at a finer interval.) (Note that in the left panel of Figure 1 values of $P(X_k)$ for $k > k'$ cannot actually be computed because of the failure of the software for such values. For display purposes these P-values have been set to an arbitrary large number.)

5.2 The sample variance

More commonly, software degrades steadily (rather than abruptly as above) in its performance for a sequence of graded data sets. The manner in which it degrades can be helpful in indicating the nature of the deficiency. An example is the sample variance (squared standard deviation discussed in Section 4) of

m numbers x_i, $i = 1, \ldots, m$, with arithmetic mean $\bar{x}$, given (stably) by

$$s^2 = \frac{1}{m-1} \sum_{i=1}^{m} (x_i - \bar{x})^2. \tag{4}$$

Some software uses the *mathematically equivalent* (but unstable) form

$$s^2 = \frac{1}{m-1} \left(\sum_{i=1}^{m} x_i^2 - m\bar{x}^2 \right), \tag{5}$$

which has the property that it suffers from cancellation of significant figures for data sets with small *coefficient of variation* $s/\bar{x}$.

Consider a basic data set defined by $X_0 = \{\mu + (i - n - 1)h, \quad i = 1, \ldots, m := 2n + 1\}$. The mean and variance of this data set are μ and $\sigma^2 = h^2(n + 1/2)(n + 1)$, respectively. A family of graded data sets can be derived from this set by adding an increasing sequence of values: $X_k = X_0 + d^k$, $k = 1, \ldots, N$, for some $d > 1$. The mean and variance of X_k are $\mu + d^k$ and σ^2, respectively. Graded data sets were so constructed using $\mu = 2.503$, $h = 0.1$, $d = 2$, $n = 12$ and $N = 30$. Software implementations of the above two formulae were applied to these data sets and performance profiles determined.

It is observed in Figure 2 that the stable algorithm (plus signs) performs completely satisfactorily for problem degrees of difficulty K spanning 10 decades. The unstable algorithm (points joined by straight-line segments) loses a number of decimal figures over and above that which would be lost by a stable algorithm in a manner essentially proportional to $\log K$. (This result is predicted by a detailed error analysis of formula (5).) A performance profile for variance software similar in behaviour to the generally rising line in Figure 2 could be taken as an indication that the unstable formula (5) had been used.

5.3 Polynomial regression

Polynomial regression software can be deficient in a number of respects. Prime causes are (i) the parametrisation of the polynomial (i.e., the choice of basis functions (monomials, Chebyshev polynomials, etc.) for representing the polynomial), (ii) the scaling or normalisation of the problem data, and (iii) the algorithm used to solve the linear algebraic equations defining the polynomial coefficients. Failure to scale or normalise the data sensibly or a poor choice of parametrisation can cause the resulting equations to be unnecessarily ill-conditioned. A poor choice of algorithm to solve this system can also unnecessarily *induce* ill-conditioning. It is to be emphasised that ill-conditioning induced in this way can have consequences so serious that some polynomial

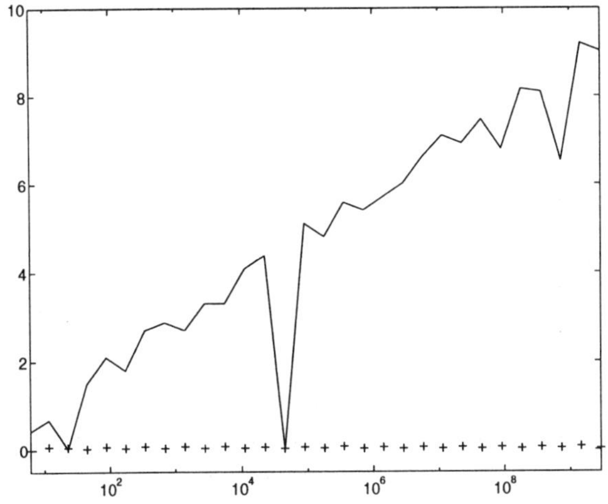

Figure 2: Performance profile for sample variance against inverse coefficient of variation calculated using a stable algorithm (plus signs) and an unstable algorithm (points joined by straight-line segments).

regression software may be incapable of producing any correct figures in the coefficients of a polynomial of degree as low as two or three for certain large sets of data. However, for these same data sets a stable algorithm is able to produce polynomial fits of very high degree (30 or more) with negligible loss of numerical precision.

A basic data set $\{(x_i,\ y_i\}_1^{21}$ was devised by generating values y_i lying exactly on a quadratic polynomial at 21 equispaced points x_i in $[-1,\ 1]$. Null-space methods were then used to perturb these values to simulate random noise. The 2-norm of the resulting perturbations is identical to that of the residuals of the quadratic fit to the perturbed points as a consequence of the null-space property. A sequence of 40 graded data sets was constructed by adding in each case $k = 1, \ldots, 40$ a constant $2^{k/4}$ to the original x_i. Thus, the data sets in the sequence are successively further removed from the origin, so simulating measurements in narrow ranges such as arise in mass and length metrology.

Three items of polynomial regression software were applied to these data sets and performance measures P based on (1) applied to the residuals of the fitted polynomial determined in each case. The resulting performance

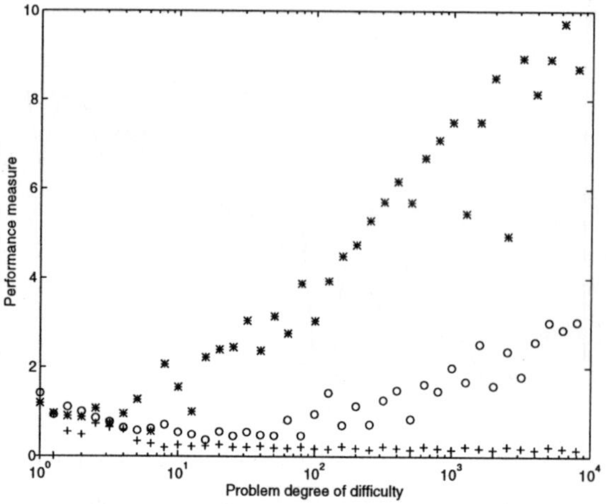

Figure 3: Performance profile for three items of polynomial-regression software when used to fit a quadratic polynomial to 21 equispaced data points. + signs: Algorithm A (normalised data, Chebyshev polynomial basis, QR decomposition); circles: Algorithm B (unnormalised data, monomial basis, QR decomposition); asterisks: Algorithm C (unnormalised data, monomial basis, formation and solution of the normal equations).

profiles (P against degree of difficulty K equal to the distance of the data set from the origin) are shown in Figure 3. The plus signs denote values obtained from the use of a stable algorithm A, which uses a Chebyshev basis in a normalised variable, and the resulting overdetermined linear system solved using QR decomposition. The circles denote values obtained from the use of a less stable algorithm B, which uses power series in an unnormalised variable, followed by QR decomposition. The asterisks denote values obtained from the use of an even less stable algorithm C, which is as B except that it forms and solves the normal equations.

Note that only Algorithm A performed near-optimally, the software based on the other algorithms losing accuracy as K increased, particularly in the case of Algorithm C.

Other dimensions of the performance of polynomial regression software that can be explored using performance profiles are polynomial degree, the distribution of x-values, and the standard deviation of the noise in the y-values. Similar behaviour is observed with regards to these dimensions of problem difficulty.

5.4 Associated circle fitting

The use of graded data sets to test software for fitting Gaussian associated circles to data in the plane is now discussed. In many instances in coordinate measurement, data is gathered lying on an arc of a circle rather than a complete circle. Since circular arc fitting is generally a difficult task, software designed primarily for fitting complete circles is often used for this purpose. The extent to which such software is appropriate can be explored using suitable graded data sets. Such sets can use for instance the angle subtended by the arc at the circle centre, or the number of equispaced points on the arc, or the standard deviation of the measurement errors as the underlying performance parameter.

Graded data sets X_k, $k = 1, \ldots, 28$, were generated as follows. Nine equispaced points lying on a circular arc of unit radius subtending an angle of $\pi/16$ radians were generated. Null-space perturbations with standard deviation 10^{-6} were made to these points, to form X_1, in such a way that the best-fitting circle to the perturbed points was the original circle. Sets X_k, $k = 2, \ldots, 28$, were generated similarly, except that the angle subtended by the points in X_k was a factor of $2^{-1/2}$ smaller than that subtended by those in X_{k-1}.

Software based on algorithms designed for full circles[8] and partial circles[9] was applied to these data sets. The resulting performance profiles are shown in Figure 4. It is to be noticed that the performance for the partial arc software (indicated by plus signs) remains consistently good as the arclength decreases. That for the full circle software is erratic for about half of the 28 data sets (small circles) and generally for those with the shorter arc lengths *fails* (indicated by asterisks, arbitrarily accorded a large performance measure value). It is to be commented that the full circle fitting algorithm[8] is excellent technology, so other circle fitting software may perform less well than this. The partial arc fitting software[9] reports correctly the natures of the failure as "failure to converge" (within 200 Gauss-Newton iterations) or "diverges", in each case of failure. Forbes[10] discusses circle and other parametrisations.

6 Concluding remarks

This paper has indicated the nature of some of the work carried out at the National Physical Laboratory on the use of reference data sets for testing software used in metrology. Particular emphasis has been placed on the use of graded reference data sets to produce performance profiles of test software. Several examples of their use and benefits have been provided.

Much work remains to be carried out before the metrologist can have suf-

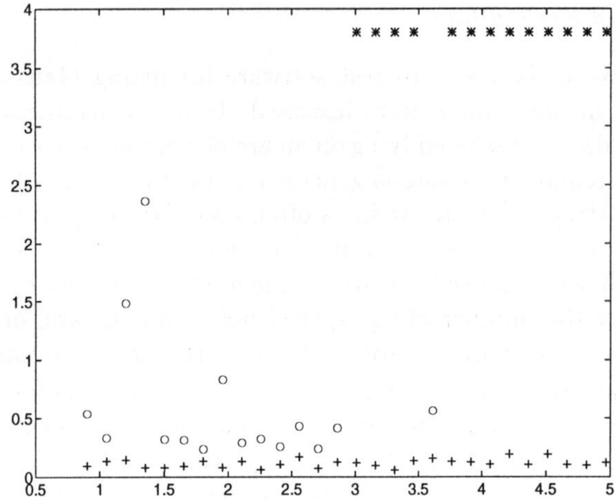

Figure 4: Performance profile for circular arc fitting software against $-\log_{10}$(length of arc) (small circles) based on a full circle fitting algorithm (small circles) and on a partial arc fitting algorithm (plus signs).

ficient confidence in the software he uses. Some of the areas that need to be addressed include (i) the further development of characterisation-based data generators, (ii) the use of hill-climbing techniques to identify weaknesses, e.g. to determine data sets which give worst-case software performance, (iii) fully automated testing including data generation and report writing, (iv) acceptance by the community of automated large-scale testing, (v) acceptance of software as an integral part of measurement traceability.

Bernard Butler provided valuable comments on a draft of this paper.

References

1. G. T. Anthony, Helen M. Anthony, M. G. Cox, and A. B. Forbes. Parametrization of geometric form. Technical Report EUR 13517 EN, Commission of the European Communities, Luxembourg, 1991.
2. G. T. Anthony, H. M. Anthony, B. Bittner, B. P. Butler, M. G. Cox, R. Drieschner, R. Elligsen, A. B. Forbes, H. Gross, S. A. Hannaby, P. M. Harris, and J. Kok. Reference software for finding Chebyshev best-fit geometric elements. *Precision Engineering*, 19:28–36, 1996.

3. B. P. Butler, M. G. Cox, S. L. R. Ellison, and W. A. Hardcastle, editors. *Statistics Software Qualification: Reference Data Sets*. Royal Society of Chemistry, London, 1996.
4. B. P. Butler, M. G. Cox, A. B. Forbes, S. A. Hannaby, and P. M. Harris. A methodology for testing classes of approximation and optimisation software. In R. F. Boisvert, editor, *The Quality of Numerical Software: Assessment and Enhancement*, pages 138–151, London, 1997. Chapman and Hall.
5. M. G. Cox and A. B. Forbes. Strategies for testing form assessment software. Technical Report DITC 211/92, National Physical Laboratory, Teddington, UK, 1992.
6. R. Drieschner, B. Bittner, R. Elligsen, and F. Waeldele. Testing coordinate measuring machine algorithms. Phase II. Technical Report BCR EUR 13417 EN, Commission of the European Communities, 1991.
7. S. L. R. Ellison, M. G. Cox, A. B. Forbes, B. P. Butler, S. A. Hannaby, P. M. Harris, and Susan M. Hodson. Development of data sets for the validation of analytical instrumentation. *J. AOAC International*, 77:777–781, 1994.
8. A. B. Forbes. Least-squares best-fit geometric elements. Technical Report DITC 140/89, National Physical Laboratory, Teddington, UK, 1989.
9. A. B. Forbes. Robust circle and sphere fitting by least squares. Technical Report DITC 153/89, National Physical Laboratory, Teddington, UK, 1989.
10. A. B. Forbes. Model parametrization. In P. Ciarlini, M. G. Cox, , F. Pavese, and D. Richter, editors, *Advanced Mathematical Tools in Metrology II*, pages 29–47, Singapore, 1996. World Scientific.
11. G. H. Golub and C. F. Van Loan. *Matrix Computations*. North Oxford Academic, Oxford, 1983.
12. IEEE Standard for Binary Floating-Point Arithmetic. ANSI/IEEE Standard. 754-1985, IEEE, New York, 1985.
13. ISO. ISO/DIS 10360-6. Geometrical Product Specifications (GPS) – Coordinate Measuring Machines (CMM). Part 6: Method for testing the computation of Gaussian associated features, 1997. International Standards Organisation proposed draft standard.
14. J. N. Lyness. Performance profiles and software evaluation. In L. D. Fosdick, editor, *Performance Evaluation of Numerical Software*, pages 51–58, Amsterdam, 1979. North-Holland.
15. W. Van Snyder. Testing functions of one and two arguments. In R. F. Boisvert, editor, *The Quality of Numerical Software: Assessment and Enhancement*, pages 155–166, London, 1997. Chapman and Hall.

Advanced Mathematical Tools in Metrology III
Edited by P. Ciarlini, M.G. Cox, F. Pavese & D. Richter

UNCERTAINTY IN CHEMICAL MEASUREMENT

STEPHEN L. R. ELLISON
Laboratory of the Government Chemist, Queens Road,
Teddington, Middlesex TW11 0LY

Current practice in uncertainty estimation in chemical measurement is briefly described, starting with an overview of chemical measurement and its applications. It is shown that formal application of the principles outlined in the ISO "Guide to the Expression of Uncertainties in Measurement" to analytical chemistry is hampered by the difficulty of obtaining good theoretical descriptions of analytical measurement processes applied to widely varying sample types. The relationship between ISO measurement uncertainty principles and ISO 5725 interlaboratory trial reproducibilities is considered. The use of reproducibility data in ISO uncertainty estimates is discussed. Two approaches are suggested; one uses a consideration of laboratory-, method- and sample-related uncertainties to provide a structure for uncertainty estimation, the other is based on reconciliation of data obtained from validation studies against a cause and effect study of the analytical process. The latter approach provides for effective use of prior data for the individual laboratory. It is suggested that wherever possible, collaborative trials should include or produce stable reference materials, both to show uncertainties associated with method bias and to allow other laboratories to determine how accurately they reproduce the trailed method.

1. Introduction

Measurement uncertainty information has always been necessary in some form to inform regulatory activity. Recently, the publication of the ISO "Guide to the Expression of Uncertainties in Measurement"[1] and derived guidance for analytical chemistry[2] have promoted a harmonised approach to uncertainty estimation. This in turn has led to increased expectations from accreditation bodies that each laboratory should determine, and if necessary report, the uncertainty associated with their own determinations.

The ISO Guide bases estimation of combined (that is, overall) uncertainties on the combination of individual contributing uncertainties expressed as standard deviations. This approach is generally applied directly to a mathematical model of the measurement process, including terms for all the factors affecting the result. An approach based on detailed consideration of individual experimental steps has been recommended as a starting point[2]. This contrasts directly with the usual approach taken in chemistry and applied by regulators, which relies on a combination of interlaboratory study data to give reproducibility figures with 'bias' information such as recovery information to convey overall uncertainties. The possibility of using such data to provide uncertainty estimates for standard methods has been suggested[2,3].

In this paper, the present state of uncertainty estimation in chemical measurement is briefly described, starting with an overview of chemical measurement and its applications. This is followed by a consideration of the applicability of ISO measurement uncertainty principles in analysis. Uncertainty estimations using collaborative study data and error propagation theory will be discussed. An example will show how a consideration of laboratory-, method- and sample-related uncertainties can provide a structure for uncertainty estimation. Some considerations for conduct of future collaborative trial are additionally suggested on the basis of experience in applying collaborative trial data for individual laboratory uncertainties. In particular, it is suggested that wherever possible, such trials should include or produce stable reference materials, both to show uncertainties associated with method bias and to allow other laboratories to determine how accurately they reproduce the trialed method.

Finally, the relationship between the two approaches will be considered. A key requirement of the ISO approach, based as it is on error propagation theory, is a clear and unambiguous model of the measurement process. It will be shown that this is by no means generally available in analytical chemistry. The problems of constructing such a model will be discussed, and alternative, experimental, approaches considered. In particular, the relationship between the ISO approach and the well established use of method validation through ruggedness testing and interlaboratory trial4 will be explored. It will be shown that, in principle, reconciliation of data obtained from validation studies against a cause and effect study of the analytical process, followed by collection of additional information, can substitute for full mathematical models of analytical experiments.

2. Chemical measurement

2.1 Chemical measurement process

A formal description of chemical measurement might be "the estimation of the amount of substance present". "Amount of substance" is formally quantified in moles, the SI unit. It follows that "substance" refers to a material composed of a single molecular or atomic species. In practice, by no means all materials of interest to chemists do consist of single molar species; for example, "fat", "protein" and most polymers consist either of a variable range of discrete substances or, for polymers, variable numbers of monomer units combined to form macromolecules of various sizes. In practice, therefore, it is at least equally common to express amounts in mass units or mass concentrations.

In general, chemical measurement involves three main steps; identification, isolation, and quantitation. The material to be quantified must be uniquely identified and distinguished from other materials in a particular mixture or 'matrix'. For analysis, the material of interest must be separated entirely from the matrix for quantification, or a signal unique to the material - such as a spectroscopic absorbtion, selective electrode voltage or other specific signal - must be measured. In practice, chemical measurement frequently combines the two; for example, a separative technique such as gas chromatography is combined with mass spectrometry to give selective mass ion detection on separated or partially separated substances. Finally, once a unique response is identified, quantification follows normal practice for the type of signal concerned. If separation is believed to be complete, the material is quantified by, for example, direct weighing. Otherwise, the instrument response is generally calibrated against reference mixtures, usually of pure substances in appropriate media. In this sequence, the crucial steps are identification and isolation. It will be seen that these, rather than the quantification step, contribute the vast majority of the uncertainty in chemical measurements.

2.2 Purpose of analysis

Chemical measurement performs a number of vital functions in society. Analysis supports regulations, such as environmental legislation and food and drug safety regulations. A substantial fraction of world trade is in minerals and chemicals, which are valued according to the amount of desired material present. Further, many goods attract customs duty according to the chemical composition. Health care, too, depends to some extent on analytical determinations for diagnosis. And of course, chemical research and development decisions are frequently based on the results of chemical measurements, both to inform choices on the direction of fundamental research and to determine optimal ranges or concentrations, as in processes or dose response. It follows that chemical measurement affects a wide variety of sectors in society, and the levels of uncertainty involved are clearly important.

2.3 Accuracy of analysis

Figure 1, after Horwitz[5], indicates the order of magnitude of uncertainties in chemical measurement as a function of the mass fraction of analyte present. The

Figure 1: Reproducibilities in interlaboratory comparisons

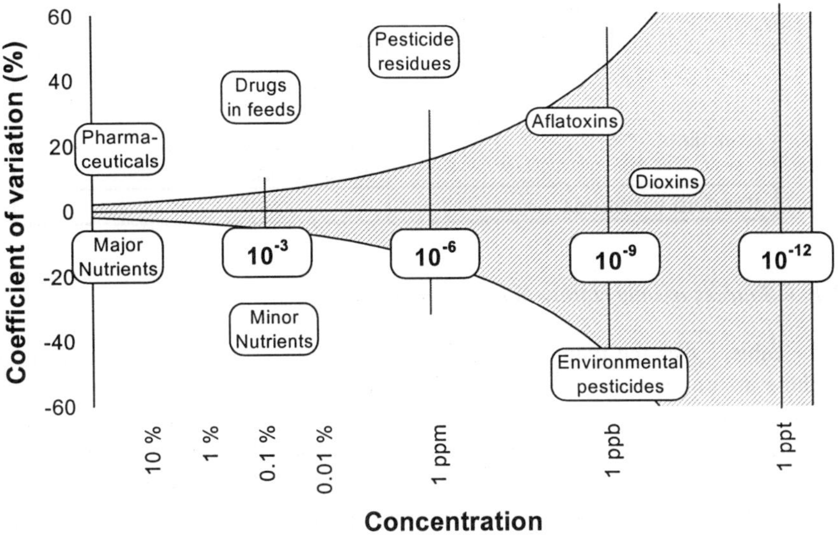

relationship shown is empirically deduced from a large number of interlaboratory comparisons, and typically predicts uncertainties within a factor of about three for routine testing. (There are, of course, many exceptions; for example, assay work can be as much as two orders of magnitude more precise). The most striking features are the size of the uncertainties, and the rapid increase at low levels of analyte. Thus, typical uncertainties in chemistry are in the percentage range, while at very low levels, the uncertainties barely permit quantification. Despite this, legislation is applied at practically all the levels shown on the diagram. Indeed, it is not uncommon that as soon as chemical measurement is capable of detecting a material at a new, lower level, legislation follows to control materials at that level. Certainly some environmental legislation based on absolute quantities of material in effluent streams becomes very hard to enforce with large effluent streams - the maximum levels for some pesticides, at 2-5g/year, are undetectable in mill effluent streams typically near 10^9 l/year[6].

With uncertainties of this magnitude, it is pertinent to ask both how the uncertainty arises, and how the uncertainty can be controlled and evaluated.

Table 1: Sources of uncertainty in chemical measurement

Incomplete definition	Sampling
Incomplete implementation	Environmental factors
Analogue readings	Instrument resolution
Calibration	Reference standards
Constants	Approximations and assumptions
Random variations	

2.4 Sources of uncertainty in chemical measurement

Table 1 lists the main sources of uncertainty in chemical measurement. Few of these are unknown in other fields of measurement; the differences are largely in relative magnitudes. In chemistry, the most important factors are incomplete definition of the measurand, in particular, whether the signal being measured accurately represents the substance of interest; assumptions about chemical or other behaviour of the analyte during the analytical process; the relevance of calibration standards (see below); sampling uncertainties; and random variability, often associated with inter-laboratory variation. Figure 2, which shows a summary of uncertainties estimated for the determination of cholesterol in fats and oils, illustrates the point; uncertainties around calibrated intermediate measures, particularly mass ('Sample wt'), reference material purity and solution volume ('Int Std') are very small; the largest single factor in this case being associated with the extraction and isolation steps. Note that in this case, using an internal (chemically dissimilar) calibration standard, the overall uncertainty is quite good by routine testing standards.

3. Estimating chemical uncertainties and the ISO *Guide*

The principles outlined in the ISO *Guide* [1] can be summarised roughly as follows:

1. Define the measurand exactly
2. Construct a mathematical model of the process relating input parameters and measurements to the final result
3. Estimate, by experiment or otherwise, the uncertainties associated with each input value, expressing each in the form of a standard deviation
4. Use first- or higher-order error propagation theory to predict the consequent uncertainty associated with the end result

Figure 2: Relative magnitudes of uncertainty contributions

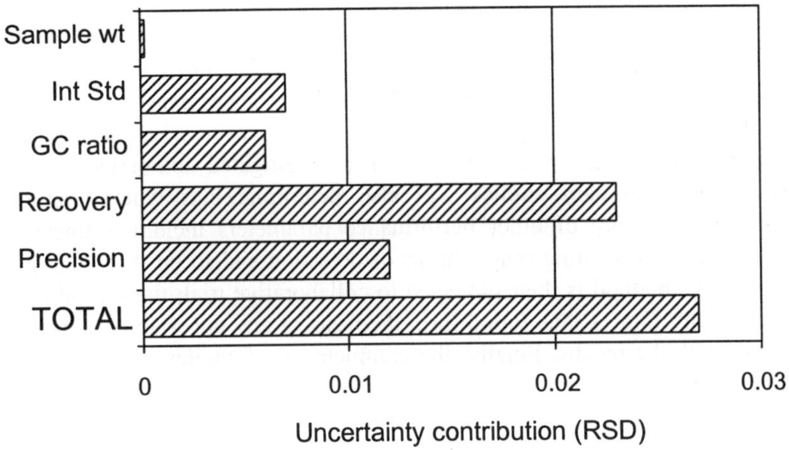

The ISO approach thus relies critically on a clear, quantitative understanding of the relationship between input effects and the result. In practice, this understanding may be expressed in the form of a complete equation, or embodied in experimental information such as the results of precision measurements or sensitivity analysis. The former may actually be practically impossible in chemical analysis, for a variety of reasons. Every determinand is unique; every different chemical element, every different molecule, every formulation can behave differently. Every 'matrix' is unique; no two substrates behave in the same way, no two chemical species interact with the substrate in the same way, and interactions between the environment and substrate are rarely understood. It follows that explicit models are very difficult to build. Nonetheless, some examples of formal application of the ISO approach have been published[2,7,8]. It is noteworthy that the largest uncertainties in these cases relate to purely chemical processes; uncertainties associated with physical measurements do not contribute significantly to the overall estimates obtained. Since the chemical processes are extremely difficult to characterise quantitatively, heavy reliance must be placed on experimental investigations. In part because of this, current practice in chemical measurement has generally taken a substantially different approach to dealing with uncertainties.

In analytical chemistry, current practice in method performance assessment relies on good method development practice and "method validation", a process by which the performance of a method is measured and compared, ultimately, with the requirements of the purpose of the measurement. During method development, procedures are established for isolation, calibration and so on. The effect of key control parameters - times, temperatures and others - is investigated. This initial process, called ruggedness or robustness testing[4], typically varies important parameters within defined limits and tests the observed changes in results for significance. Note that no attempt is usually made to determine the actual rate of change of the result with changes in the parameter. However, where a parameter is found to be affecting the result, the reason is investigated and/or closer control adopted. The preliminary studies will also determine bias where applicable, together with a variety of other performance parameters including linearity of response, practical operating range, limits of detection and so on. Provided all are satisfactory, the method is then subjected to collaborative trial; the reproducibility and repeatability figures thus found form the basis of any uncertainty conveyed in association with the results. Finally, the complete set of studies may be reviewed for fitness for purpose.

4. Using collaborative trial data for uncertainty estimation

Given a wealth of data acquired from method validation, it is reasonable to ask whether, and how, such data can be used to form ISO uncertainty estimates. Clearly, use of such an estimate to inform an uncertainty estimate for an individual laboratory result constitutes a classic type B estimate, in that it is not derived from repeated measurements on the system in question. Such estimates are explicitly accepted within ISO principles. It remains to determine how such data can be applied in practice

One approach suggested for using reproducibility data in uncertainty estimation uses a simple model involving a method bias contribution, a sample contribution, and a laboratory performance contribution[2]. The results of such a study are shown graphically in Figure 3. The example involves analysis of a foodstuff for meat content. Fat is extracted and determined gravimetrically, and a simple reproducibility standard deviation is used as a working estimate of uncertainty. Total nitrogen is determined by Kjeldahl analysis, and converted to estimated protein content via an appropriate 'Nitrogen Factor'. The contributions (shown in the form of relative standard deviations) are estimated as follows. The Nitrogen factor is specific to beef, but varies with the particular 'cut' of meat. The single factor used is an average; however, published material shows a range of values from which a standard deviation can be obtained. This contribution is shown as NF in Figure 3, and is considered the main contribution arising from sample effects. Assuming that actual nitrogen content (rather than nitrogen defined as 'Kjeldahl nitrogen') is required, it is possible that the method is biased;

Figure 3: Contributions to uncertainty in meat analysis

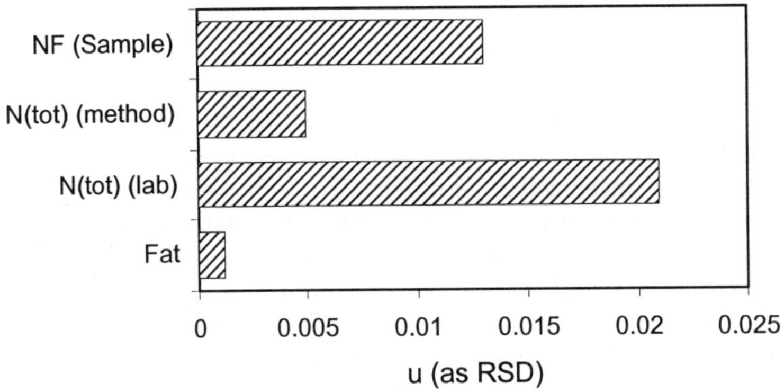

comparisons of extensive studies of Kjeldahl determinations with combustion methods suggest an uncertainty contribution from this source of 0.012 (as RSD). This is associated with total nitrogen, and is considered the main uncertainty contribution associated with the method itself. Finally, the laboratory performance (combined bias and precision) is estimated from reproducibility studies and appears as a second contribution to the uncertainty in nitrogen content N(tot). In this instance, additional information from continued inter-laboratory proficiency assessment confirms the applicability of the latter estimate.

This simple approach allows a reasonably quick estimate of uncertainty from available data. It does, however, have shortcomings compared with uncertainties derived from input measurements traceable to international standards. In particular, there is no clear link between the control applied to operational parameters for the method and the final uncertainty. More generally, without additional intercomparison work, it is impossible for a laboratory to confirm that its results are indeed within the range obtained in the collaborative trial of a method. This points to a simple possible improvement of collaborative trial practice; at least one, and preferably more, reference materials should be studied in the trial, so that laboratories not included in the initial trial can compare results directly with the same reference material. Ideally, a reference material would be used as a mandatory check on the procedure, though in practice the availability and limited lifetime of reference materials makes this approach hard to apply.

5. Cause and effect analysis

To overcome these and other difficulties, it seems sensible to attempt reconciliation through more detailed study of the analytical method. Table 2 compares the method validation and ISO measurement uncertainty approaches directly. There are important similarities; in particular, both require an identification of the main factors affecting the result, and both evaluate each parameter's effect and make an overall estimate of uncertainty. The differences are, however, striking. No formal model of the measurement process is drawn up in the validation process. The effect of individual parameters is not measured, only tested for significance compared to overall precision. The final 'uncertainty' is estimated in quite different ways; the independently assessed reproducibility forms the best estimate of uncertainty in the 'validation' approach. The latter, interestingly, is based on a basic error model in which 'within run' and 'between laboratory' effects are estimated, implying that the laboratory and the analyst are critical input parameters; other critical factors for chemistry may be the (ill-defined) nature of the sample matrix and the amount of material present. These differences make it hard to see how the two can be reconciled. However, I now describe a procedure which we have developed in our laboratory and which offers a degree of reconciliation through a process we term 'cause and effect analysis', after the visual tools used.

Cause and effect analysis, as we apply it, is based on the observation that correctly planned experiments can obtain most of the information required to assess uncertainty. To plan experiments properly, we apply the initial ISO principle of identifying the main effects on the final result, in the form of a list of input effect.

Table 2: Comparison of method validation procedures with ISO uncertainty estimation

ISO Uncertainty	Method validation
Identify main effects on result	Identify main effects on result
Quantify uncertainties in values	Set appropriate control ranges
Quantify contribution to result	Vary input parameters
	Test observed changes for significance
Combine effects	Determine reproducibility independently

We structure that list in the form of a cause and effect diagram (sometimes known as an Ishikawa or 'fishbone' diagram); the basic principles of the construction are detailed elsewhere[9]. The basic strategy is:

- Write the complete equation for the result. The parameters form the main branches of the diagram
- Consider each step of the method and add any further factors to the diagram. Examples include temperature and matrix effects
- For each branch, add contributory factors until effects become sufficiently remote
- Resolve duplications and re-arrange to clarify contributions and group related causes

Figure 4 shows a typical, slightly simplified cause and effect diagram (omitting a precision branch) for the cholesterol determination discussed earlier. The main 'branches' relate to individual parameters in the expression for the result (concentrations, volumes, weights etc.); the subsidiary branches represent contributory effects on the result.

Aside from the clear, visual structure of the factors affecting the result and their interrelationships, which is useful in itself, the structured list has an additional value. It can, with care, be used to plan suitable experiments and identify requirements for additional information. Given existing experimental data (including collaborative trial data) it is possible to reconcile the data with the information required, to assess the need for further experiment or information. In planning additional experiments, we use the fact that where an input parameter has been varied representatively during an experiment, the observed variation in the result includes the contribution to overall uncertainty from that parameter. In this context, *representative variation* means that the variation in the particular factor must cover either the full range permitted by the method or, (where there are no control limits specified), the full range generally covered in practical implementation of the method, and with a similar distribution. Since in most practical experiments, many parameters vary simultaneously, and will often naturally vary representatively given experiments covering a range of time, analyte levels, matrices and so on, it will often be possible to design experiments, or use data, that provide a single estimate covering the majority of the uncertainty. In principle, therefore, it is possible to carry out an analysis showing exactly the extent to which any method validation work has covered the variation required to obtain an uncertainty estimate fully compliant with ISO principles, and where additional information is needed.

Figure 4 shows just such a reconciliation. Given an experiment in which the actual recovery of material is determined from a wide variety of materials (covering the full scope of the method), and with sufficient replication to include

Figure 4: Cause and effect diagram for cholesterol determination

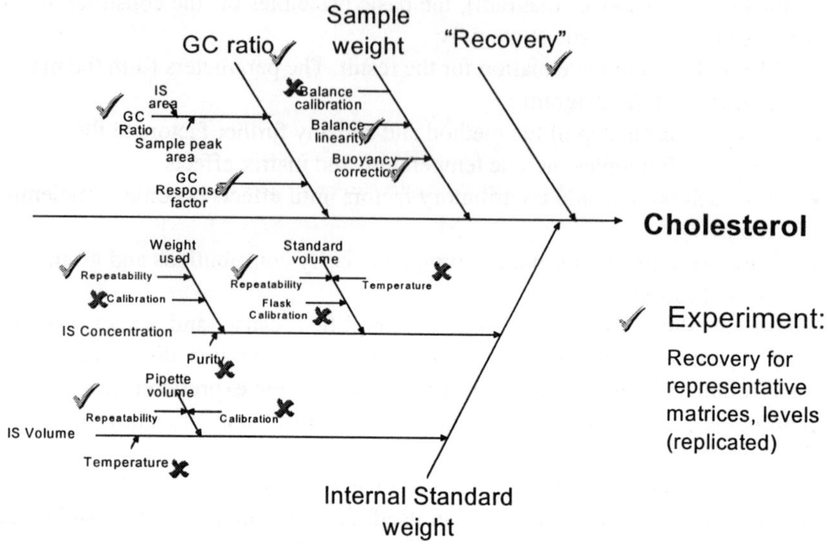

all run to run precision terms, it is possible to 'tick' all the factors whose contributions will appear within the overall distribution of results. The remaining factors, identified by crosses in Figure 4, are in this case chiefly dependent on calibration information, and the relevant information will be to hand in any accredited laboratory. The associated uncertainties can then be combined with the existing estimate following standard ISO principles (or, if genuinely negligible, simply recorded and neglected). Note that the 'recovery' branch is naturally more complex; in determining recovery of material, it is of course necessary to include the uncertainties in the amount believed to be present and make due allowance for any other contributing factors.

Of course, in many applications, complete experiments will not be practicable and substantial contributions will not have been included. This is typical for 'in-house' methods. In those cases, the available experimental data will not provide as much of the information. The cause and effect analysis will still show where additional information is needed, and the information will need to be sought in other ways. Here, the ISO approach becomes particularly relevant; modelling, prior information and other tools will be employed to provide type B estimates of individual components. The approach therefore represents an effective general methodology reconciling current analytical practice with ISO principles.

6. Summary

The sources and sizes of uncertainties in analytical chemistry have been discussed in detail, and a summary of approaches to achieving intercomparability has been presented. Current practice in assessing uncertainty in chemistry has also been described. This has been compared with the recommendations in the ISO *Guide to the expression of uncertainty in measurement.* There are substantial differences in both philosophy and detailed approach. This follows in part from the lack of good theoretical models for analytical measurement processes, and in part from a long and largely successful history of reliance on standardised methods validated by collaborative trial. However, there are sufficient similarities to indicate that the same basic information is being collected. In this paper, a general approach to reconciling the two different methodologies has been suggested. The cause and effect analysis described provides a good picture of all the important effects; careful study allows the analyst to check the applicability of existing data; and the ISO guide gives principles which show how to combine the relevant data with additional information.

Acknowledgement

Production of this paper was supported under contract with the Department of Trade and Industry as part of the National Measurement System Valid Analytical Measurement Programme.

References

1. *Guide To The Expression Of Uncertainty In Measurement.* ISO, Geneva, Switzerland 1993. (ISBN 92-67-10188-9)
2. *Quantifying uncertainty in analytical measurement,* Laboratory of the Government Chemist, London (1995) (ISBN 0-948926-08-2).
3. Analytical Methods Committee, *Analyst* **120** 2303 (1995)
4. W. J. Youden, E. H. Steiner, *Statistical manual of the Association of Official Analytical Chemists*, AOAC, Arlington VA (1994). ISBN 0-935584-15-3
5. W. Horwitz, *Anal. Chem.* (1982)
6. S. L. R. Ellison, A Williams, V J Barwick, *ESARDA Bulletin* (1996)
7. A. Williams, *Anal. Proc.* **30** 248-250 (1993)
8. M. Pueyo, J. Obiols, E. Vilalta, *Anal. Commun.* **33** 205-28 (1996)
9. ISO 9004-4:1993, *Total Quality Management Part 2. Guidelines for quality improvement*, ISO 1993

Advanced Mathematical Tools in Metrology III
Edited by P. Ciarlini, M.G. Cox, F. Pavese & D. Richter

Mathematical Methods for Data Analysis in Medical Applications

J. HONERKAMP
Fakultät für Physik, Albert-Ludwigs-Universität, 79104 Freiburg, Germany

A general framework is presented which can often be applied in data analysis: the data can be interpreted as the observation of a hidden dynamical system. The model then includes an observation equation in addition to the system equations. By reporting about three projects, on which we work in collaboration with medical scientists, this framework will be outlined. The different mathematical problems which arise and the methods for solving them are shortly discussed.

1 Hidden dynamical systems in data analysis

The number and complexity of physical measurements in modern medicine is growing from day to day. More and more data are produced which cannot be analyzed and interpreted sufficiently by counting and visualizing alone. Therefore, a data analysis with modern methods from mathematical statistics and stochastics will play an increasing role in medicine.

We can distinguish two classes of data analysis:

- A model-free analysis is more or less only an exploration of the data with the objective of a classification. Only a few assumptions are needed for such an analysis. Of course, every diagnosis of a physician is a classification and measuring the feaver is already an example of a physical measurement which leads to a classification. There are many more sophisticated measurements and more difficult tasks of classification. But I will not discuss them here.
- In a model-based analysis the input of assumptions is much larger. Therefore the results of an analysis can also be much more detailed. The spectrum of mathematical methods which are needed is also much richer. My examples will be of this type and I will show that a model-based analysis with efficient mathematical methods can lead to a progress in our knowledge of physiological processes and to quantitative measures for the effect of a medical treatment.

Before I go into the details of specific applications let me present the general structure of the models we consider. In our experience this structure can be identified in every model based analysis. A complete model includes

- a model for a quantity X, which is in general not directly accessible. This quantity may depend on space and time (or on any other quantity). It may stand for the density of a tissue, the rate of a flow, the concentration of a substance, etc. The equation for this quantity may be

- a stochastic equation such as a Langevin-equation or a master equation for the density $p(x)$ of X
- a deterministic differential equation for $X(t)$ or $X(r, t)$
- an assumption about the dependence in space as e.g. in the case of Markov Random Fields.

I mentioned that the quantity X is not directly accessible. Therefore a model for this system variable is not enough, we need also

- an observation equation

$$Y = C[X] + \epsilon ,$$

which relates the observable quantity Y (identified with the measured quantity) to the system quantity X by the functional C. ϵ is a random variable which takes into account that every measurement is affected by noise.

Thus we distinguish between the observation Y and the quantity X for which a model can be formulated which is simple enough. This is also reflected in the well known models of mathematical stochastics such as the State Space Model or the Hidden Markov Model.

Very important for all measurements in medicine is that these are as little invasive as possible. Thus we find in medical applications especially many methods in which the relation between the quantity of interest and the quantity which can be measured is very indirect. Therefore this concept of hidden dynamical systems is especially adaequate for medical applications. The inference from the signals to the quantity of interest is, however, often a considerable and challenging task.

2 Determination of the permeability of the blood-brain-barrier

Collaboration: W.Huber, K.Kopitzki, J. Timmer (FDM) and
P.C. Warnke (Abt.Stereotaktische Neurochirurgie, Neurozentrum, Freiburg)

In healthy persons the border between the brain tissue and blood vessels is impenetrable to water soluble molecules. This strict demarcation is very important for the protection of the brain from substances it cannot tolerate.

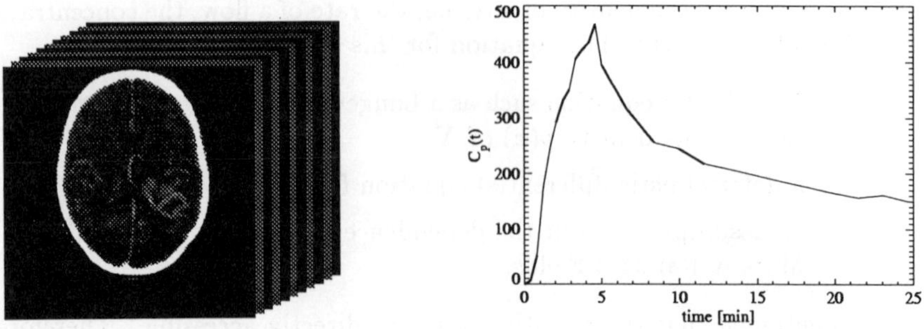

Figure 1: Typical results of the measurement of the amount of contrast medium $A_m(x,t)$ in course of time t (left) and of the concentration $C_p(t)$ of the contrast medium in the blood system (right).

The border is called the blood brain barrier. For persons with diseases like cancer the barrier is defect.

In a medical treatment of cancer the permeability can, however, also be used to bring the drugs into the tumor tissue. In order to do this under control one has to know the permeability. Only then it is possible to decide whether a chemotherapy is resonable at all, and if it is, to estimate which dose is optimal. Furthermore, one would like to know to what extent the permeability can be changed by specific drugs.

For the measurement of the permeability, a contrast medium is infused into the patient with constant rate over five minutes. This marker will spread out very fast to the whole blood system and eventually penetrates through the blood brain barrier into the brain tissue. By a computer-assisted tomography one can measure the amount of contrast medium in small volume elements of the brain. In (fig. 1, left) results of such measurements are shown. We will call the value of this amount $A_m(x,t)$, where x runs over the volume elements (voxels). This is not the only information obtainable in this experiment. One can also measure the concentration of the contrast medium in the blood plasma. We will call this $C_p(t)$. A typical time series of $C_p(t)$ is shown in (fig. 1, right).

The problem now is to infer from this information the permability of the blood brain barrier in each volume element. A model for the exchange of a medium between blood and brain tissue, which can be found in the literature[1] is the following:

We consider a two compartment system (fig. 2), the compartments are the blood system and the brain tissue respectively.

Let us call the amount of contrast medium in the brain tissue (extravas-

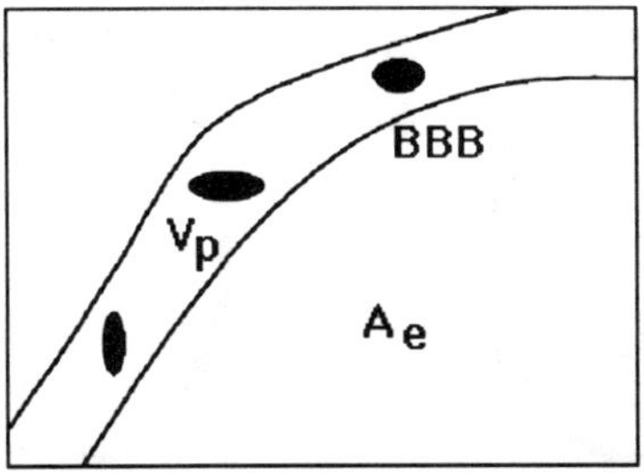

Figure 2: The brain tissue and the vascular system can be regarded as two compartments which are separated by the blood brain barrier (BBB).

cular region of a volume element) $A_e(x,t)$. If V_p is the amount of plasma in a volume element, then $V_p\, C_p(t)$ represents the amount in the blood system and we have, adding up

$$A_m(x,t) = V_p(x)C_p(t) + A_e(x,t) + \epsilon(x,t) \ , \tag{1}$$

where $\epsilon(x,t)$ takes into account that the measurement is affected by experimentell errors. Furthermore, let $k_1(x)$ be the transfer rate from the blood system to the brain tissue and $k_2(x)$ be the rate of transfer in the opposite direction. Then the simple equation

$$\frac{\partial}{\partial t}A_e(x,t) = k_1(x)V_p(x)C_p(t) - k_2(x)A_e(x,t) \tag{2}$$

can be formulated. There is an equation for $C_p(t)$ too, which, however, we do not need because we are provided with measurements of this quantity (see fig. 1, right).

Eq (1) and (2) constitute our model. This is very simple until now. We have again an equation for the variable of a hidden system. This is equation (2) for $A_e(x,t)$. And we have in (1) an observation equation which relates this system variable $A_e(x,t)$ to the measured quantities $A_m(x,t)$ and $C_p(t)$. The quantities we would like to estimate are $k_1(x)$, $k_2(x)$ and $V_p(x)$.

Hence, we translated the medical problem into a mathematical one: The estimation of parameters in a state space model given by eq. (1) und (2).

Because the model equations are so simple and mathematically harmless we can use the least squares estimation method to infer the rates $k_1(x), k_2(x)$ and $V_p(x)$ from the data. The only problem is that we have so many equations (one for each volume element and we have 512×512 volume elements). Therefore, to save calculation time, we follow a three stage strategy: We start

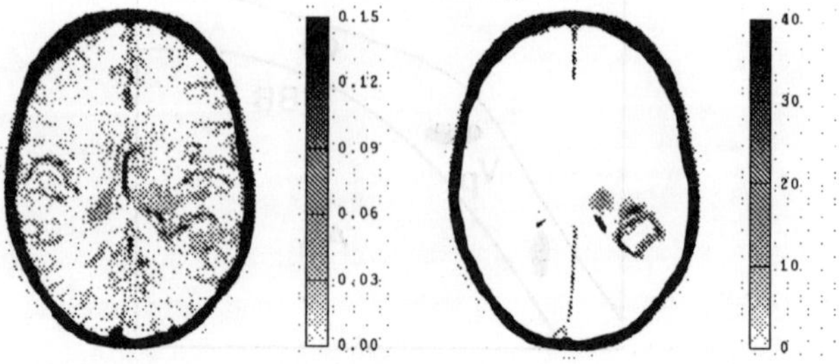

Figure 3: Results of the estimation of $V_p(x)$ (left) and $k_1(x)$ (right).

with the assumption $V_p(x) = 0, k_1(x) = 0, k_2(x) = 0$ in each region and test whether this is compatible with the data in that region. In the volume elements where this hypothesis is not accepted we formulate a weaker assumption: $V_p \neq 0, k_1(x) = 0, k_2(x) = 0$. This we test again. Only in those volume elements, in which this hypothesis is also not accepted, we assume that the full model applies and we estimate the nonzero values of $k_1(x), k_2(x)$.

Fig. 3 shows some results. In the tumor-region there is indeed a significant permeability. A detailed analysis of the results reveals that there are strong variations of permeability in the tumor tissue and that there is no correlation of permeability and blood supply. Furthermore, the permeability is recognized as the limiting factor for an adaequate enrichment of the tumor tissue with a drug. A further benefit will be that one has obtained a quantitative measure for the effects of a drug or a therapy.

Until now we have neglected the diffusion of contrast medium between the volume elements. On a larger time scale (several hours) this will become relevant and we have to take it into consideration in our model. Then our mathematical problem becomes a problem of the estimation of parameters in hidden partial differential equations. This is a much more involved problem and we will see which difficulties arise.

This method to estimate the permeability of the blood brain barrier developped so far is used already in the hospital of the University of Freiburg for testing some drug.

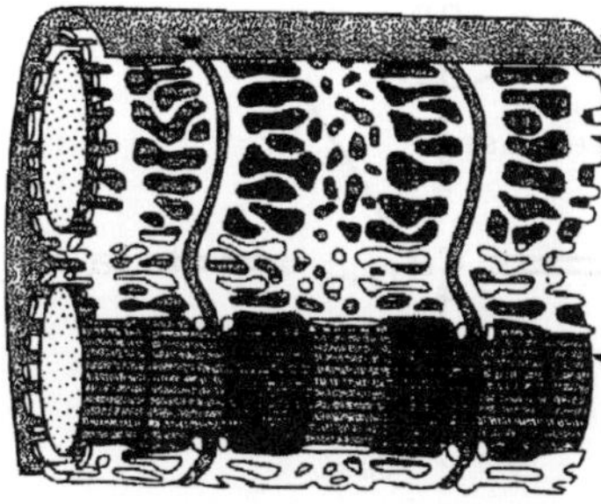

Figure 4: Compartments of a muscle cell (taken from: B. Alberts, D. Bray, J. Lewis, M. Raff, K. Roberts and G. Watson, The Molecular Biology of the Cell. New York: Garland Publ. 3rd ed. 1994.)

3 The release of Ca^{2+} from the sarcoplasmic reticulum in a muscle cell

Collaboration: T. Müller, J. Timmer (FDM) and
W. Melzer (Abt. für Angewandte Physiologie, Universität Ulm)

The compartments of a muscle cell which are relevant for the following discussion are the cell membrane, the contractile filaments, the sarcoplasmic reticulum and the intracellular fluid between the sarcoplasmic reticulum (SR) and the filaments (see fig. 4).

In a resting state of the cell there is a low concentration of Ca^{2+} in the intracellular fluid. If, however, the cell membrane is depolarized after a nerve impulse, the sarcoplasmic reticulum is caused to release a relatively high amount of Ca^{2+} in a short time of several milliseconds so that the concentration of these ions in the intracellular fluid strongly increases. This leads to a contraction of the filaments.

Hence, the contraction of a muscle is triggered by the Ca^{2+} release of the sarcoplasmic reticulum. This release is nonzero only during the short time of the depolarisation, it can be quantified by a time dependent release rate $\dot{Ca}(t)$ which we take as constant throughout the whole intracellular fluid.

The idea now is that the specific form of this time dependence is a characteristic feature for the condition of the cell and the release channels. In certain diseases, for instance malignant hyperthermia, the release rate may be altered.

The concentration of the Ca^{2+} in the intracellular fluid cannot be measured directly. One adds a dye to the fluid, which reacts with Ca^{2+}. One can measure the concentration of the dye-Ca^{2+} complex, and the problem is to

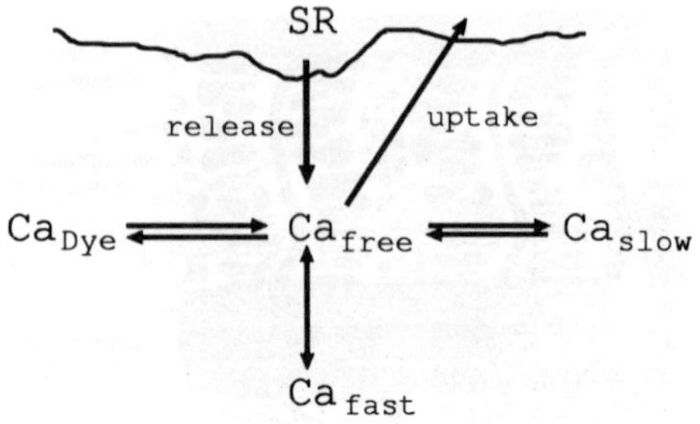

Figure 5: The scheme of reactions in the intracellular fluid.

infer from such signals the time dependence of the release rate.

This is a delicate inverse problem. Besides the dye-Ca^{2+} complex there are many other substances in the intracellular fluid which react with the Ca^{2+}-ions, i.e. which bind Ca^{2+} with rapid or slow kinetics (Ca_{fast}, Ca_{slow}). Especially there is a pump, a physiological mechanism, which transports Ca^{2+} ions back to the sarcoplasmic reticulum. Thus even a crude approximation to the real situation in the cell leads to a fairly complicated reaction scheme[2] (see fig. (5)). The rates of all these reactions are not known exactly, nor can one measure the concentration of the various substances. The only measurable quantity is the concentration of the dye-Ca^{2+} complex Ca_{Dye}.

The situation seems to be hopeless, but it is not. The important observation is that there are two time regimes. In the first regime, during the depolarization, the Ca^{2+} release $\dot{Ca}(t)$ is nonzero. This function of time could be estimated from the dye concentration signals if all other reaction rates in the reaction scheme were known. Thus, one must find a strategy to estimate these rates first.

This is just what can be done in the second regime which starts at the end of depolarization, after the release has ceased and the Ca^{2+} concentration in the fluid relaxes to its value before the depolarization. Therefore we have to formulate a state space model in both regimes.

The observation equation will be the same in both models. The systems of differential equations for the concentrations of the substances which are not directly observable differ in a small but very crucial point: One system contains the time dependent release rate $\dot{Ca}(t)$, the other not.

In the following I will concentrate on the task which has to be done first: The estimation of the reaction rates from the dye data in the second regime. (The subsequent task, to infer the release rate from the data in the first regime, given the reaction rates, is relatively simple.)

This problem is again one, in which the parameters of a differential equation have to be estimated from observations. This is the same type of problem as in the preceding paragraph, but now the system of differential equations is much larger and therefore also the dimension of the space of parameters and initial conditions. A minimization of an objective function based on least squares becomes very difficult because this space can contain many local minima or even regions of parameters and initial conditions which correspond to diverging solutions.

There exists, however, a more sophisticated and useful method for this task. Bock[3] has applied the multiple shooting approach to the problem of estimating parameters in differential equations. The idea is the following: Usually one starts a trajectory at $t = 0$ with given parameters and initial conditions, solves the system for the whole time interval and then tries to adapt these parameters and initial conditions so that consistency with the data is achieved. In the multiple shooting method, such trajectories are started at several times. In general, these will lead only to a piecewise continuous trajectory over the whole time region. The initial conditions for the system variables and the coefficients of the differential equation then have to be adapted so that finally a continuous trajectory emerges which is consistent with the data over the whole time interval. In general, one needs many data sets in order to be able to estimate the parameters and the release rate with sufficient reliability.

Different models are offered in the literature for the reactions in the intracellular fluid. The common objective is to reveal the mechanism of the calcium release. We can compare these models with respect to their efficiency for this estimation in order to find out the simplest and fastest method which can be used for the characterization of Ca^{2+} release mechanism in the cell.

The kinetic of ion channels in cell membranes

Collaboration: S.Michalek, M.Wagner, J.Timmer (FDM) and
M.Kohlhardt (Physiologisches Institut der Universität Freiburg)

Cells change their concentration of ions with the help of large proteins in their membrane. These proteins can be in different conformations, some of which conduct ions through the membrane. Then, the protein can be regarded as an open channel for specific ions (fig. 6). The protein can change its state spontaneously, the dwell time in a given state seems to be a random variable with a certain probability density.

This random walk of an ion channel through its states can be seen in experiments where one can measure the current of ions flowing through just

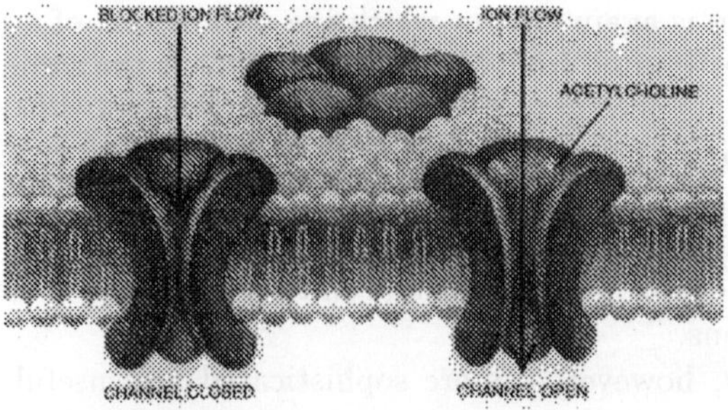

Figure 6: Ion channels in a cell membrane are proteins, which can be in different conformations (figure taken from E.Neher, B.Sakmann: The patch clamp technique, Scientific American, 66(3), 28-35 (1992))

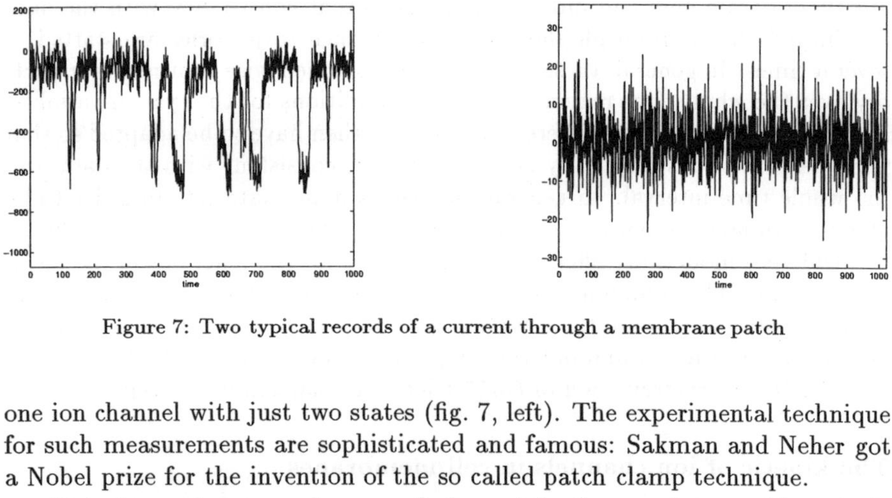

Figure 7: Two typical records of a current through a membrane patch

one ion channel with just two states (fig. 7, left). The experimental technique for such measurements are sophisticated and famous: Sakman and Neher got a Nobel prize for the invention of the so called patch clamp technique.

Not always, however, does one find a patch of a membrane with a single channel, and not always does the channel possess only two states. Thus an experimental record of a current through a patch can also look like fig. 7, right. The fluctuation of the current which is impressing already in fig. 7, left, can be so large that the conductivity levels of the channels cannot be seen any more. What one likes to learn from such experiments can be formulated with the following questions:

- How many states has a specific type of channels?
- How can one characterize the kinetics of the channel? Between which states do transitions occur and what are the rates for this?

- What is the conductivity of a type of channel in each state?

Obviously the parameters in this characterization will depend on various external conditions such as

- the concentration of ligands in the extracellular fluid
- the electric field
- the mechanical stress applied to the membrane.

For this problem the analysis with a model of a hidden stochastic process seems to be especially adaequate. Since the transitions from state to state occur spontaneously, a Markov model can be used. But the states cannot be measured directly, only the noisy current through the whole patch. An observation equation may describe this with a relation between the measured current and the state of the channel in the patch. Thus we have to formulate a Hidden Markov Model which contains all the parameters which we want to know. From the record of the current we have to infer these parameters.

It has been shown that the typical maximum likelihood estimator can be used for this inference (see e.g. Becker et.al.[4]). But there are some technical difficulties. The most interesting one results from the fact that the transition probabilities (which are closely connected to the rates and which are estimated more directly) lie either very near to 1 or near to 0. The finite size behavior has to be studied in order to find the typical sample size for which the asymptotic region with its Gaussian behavior is reached.

Once one knows how to identify the parameters of the model in a given Hidden Markov Model on the basis of a given current record, one has to check whether the proposed model with its assumptions about the number of states and number of channels in the patch can be regarded as consistent with the data. We have to test whether there are several consistent models. This is often the case. Then we have to find out by a further statistical test which model is the most appropriate one.

In the same way as one has to study how the maximum likelihood estimator works the properties of the test quantities have to be elaborated to judge the power of the test.

With all these methods at hand and using them reliably one can answer many questions concerning the ion channels by analyzing data from patch clamp experiments. One may try to reveal the gating scheme i.e. the rates and conductivity levels of a specific channel and, furthermore, how the gating scheme is changed with the external conditions. One could also be able to decide whether the gating scheme of a mutant of an ion channel differ from that of the wild type. Furthermore, one can test e.g. whether the channels in a patch do function really independently or whether the channels in a patch are all of the same type.

Data analysis of path clamp data is important for understanding the functioning of ion channels. We think that hidden Markov models will become an efficient and standard tool in the future for those data which are very noisy and therefore cannot be analyzed by simpler methods.

Conclusions and Outlook

I have reported on three projects in which the data from medical laboratories have been analyzed in terms of a hidden dynamical system. In all projects medical scientists and physicists work closely together. The aim of such a project is always a well tested and reliable software tool for the medical laboratory with a user friendly interface.

The concept of a hidden dynamical system is very general and far reaching and I believe that it finds application not only in the problems I reported but in many other problems too. I have discussed two problems, where the equation for the only indirectly observable system can be taken as deterministic and one problem where those equations have to be stochastic. In each problem we needed a different mathematical method. This is my general impression after some years of experience with data analysis of real data: The spectrum of methods to be used has to be very broad in order to be able to deal with all aspects of the data and the medical context.

References

1. D.R. Groothuis, et.al, Ann.Neurology **30**, 581 (1991).
2. W. Melzer, et.al, J.Physiol. (London) **372**, 261 (1986).
3. H.G. Bock in *Springer Series Chemical Physics, 18,* ed. Ebert, Deuflhard, Jäger, 102 (1981).
4. J. Becker, J. Honerkamp, J. Hirsch, E. Schlatter, R. Greger, Pflügers Archiv **224**, 328 (1994)

see also:
http://phym1.physik.uni-freiburg.de/~hon

Advanced Mathematical Tools in Metrology III
Edited by P. Ciarlini, M.G. Cox, F. Pavese & D. Richter

HIGH-DIMENSIONAL EMPIRICAL LINEAR PREDICTION

HUNG-KUNG LIU

National Institute for Standards and Technology, Gaithersburg MA, USA

Many engineering problems involve high-dimensional observations with mean vectors sitting in a lower dimensional space. Exhaustive measurement of all the elements of an observation is often time consuming and expensive. Applying a traditional multivariate linear model, one can incorporate a small number of the elements of the observation with a known design matrix to predict the rest of the elements. However, for a complicated engineering system, the design matrix is often hard to be fully determined. In this paper, we investigate an empirical linear model, in which we allow ourselves to use the data to determine the size of the design matrix and to estimate the unknown part of the design matrix. This estimated model is then used to construct point and interval estimation for the future observation. This technique is called HELP (High-dimensional Empirical Linear Prediction).

1 Introduction and Notations

In the industrial manufacturing of many complex products, quality assurance for each product requires making a large number, say m, of measurements to determine conformance to specifications. For a given product, y denotes its *discrepancy vector*, i.e., the $m-$dimensional vector whose components are the differences between the observed measurements and the targeted specifications. The ideal product with no measurement error therefore corresponds to a discrepancy vector which is the zero vector. Exhaustive checking of all test points for each product is often time consuming and expensive. However, the behavior of a product often depends only on a small number of independent variables. By using a Taylor expansion, we assume the model

$$y = \mu + \chi\beta + X\gamma + \varepsilon. \tag{1}$$

Here β and γ are respectively the k and ℓ dimensional unknown column vectors. Also the $m \times l$ matrix χ represents the part of the design that is unknown, and the $m \times l$ matrix X represents the part of the design that is known. Model (1) is therefore called the *empirical* linear model since part of the design matrix χ can only be determined empirically.

To make χ and β identifiable, we first write χ as

$$\chi = (S, S_0) \begin{pmatrix} D \\ 0 \end{pmatrix} V = SDV \tag{2}$$

by the singular value decomposition. Here (S, S_0) and V are orthogonal square matrices of sizes m and k respectively. It is assumed that S has size $m \times k$ and $k < m$. The matrix D is a diagonal matrix with positive diagonal elements d_i which decreases in i. Since we shall assume that d_i's are different, $d_1 > d_2 \ldots > d_k$. Throughout the paper 0 represents the zero matrix of appropriate size which is $(m-k) \times k$ in this case. Also throughout the paper, we shall follow the rule below in doing a singular value decomposition:

the first non-zero element of each column in (S, S_0) is positive. (3)

This, together with the assumptions that d_i's are distinct, implies that the columns of S are uniquely determined. Model (1), after substituting χ by (2) and replacing $DV\beta$ by η, becomes

$$y = \mu + S\eta + X\gamma + \varepsilon, . \tag{4}$$

An empirical liner model provides a useful structure in modeling a complicated engineering system when the design matrix is hard to be fully determined. The question we want to address in this paper is: can conformance to specifications be determined by a t–dimensional subvector y_1 of y?

If S were known, applying a traditional multivariate linear model, one can incorporate y_1 with the known design matrix to predict the rest of the elements. Since S is unknown, it will be estimated by using a set of y^i's, called the training data. We can express the training data from the same manufacturing process as

$$y^i = \mu + S\eta^i + X\gamma^i + \varepsilon^i, \quad i = 1, \ldots, n. \tag{5}$$

Then the estimated model, together with a t–dimensional subvector y_1 of y, may be used to estimate the rest of the components y_2 of y.

What are the reasonable point predictors which allow us to do so? How does one assess the prediction error? These are the central questions that we will address in this paper. We tackle the problems (especially the second one) by constructing prediction sets, confidence sets and Scheffé type (1959) simultaneous intervals. This prediction based on the empirical linear model is called the high–dimensional empirical linear prediction (HELP). HELP enables the engineers to test each new product at only a small number of test points and still provide a reliable assessment of conformance to specifications.

This is a challenging problem involving several interesting and difficult areas such as estimation of dimensionality, variable selection, calibration and prediction. Another striking feature is that the model could involve a design matrix that is totally unknown. In fact, the special case of (4) and (5),

$$y = S\eta + \varepsilon, \text{ and} \tag{6}$$

$$y^i = S\eta^i + \varepsilon^i, \tag{7}$$

i.e., $\mu = 0$ and $l = 0$, corresponds to the case when the design matrix is totally unknown. To emphasize the difference, we shall call model (6) *purely* empirical linear model, and model (1) or (4) *partially* empirical linear model.

2 Review of Existing Results

We now summarize some useful known results in carrying out HELP using a purely empirical linear model. Results for the partially empirical linear model can similarly be stated.

Prediction Set with Known S.
We shall assume without loss of generality that the observed elements of y, denoted by a t–dimensional column vector y_1, are from the first t coordinates. Hence we may write $y' = (y_1', y_2')$, where y_2 consists of the unobserved elements of y. In practice, one chooses the coordinates of y_1 using algorithms described below. We shall partition S as $(S_1', S_2')'$, where S_1 consists of the first t rows of S. If S were known, a standard prediction set based on y_1 is described by the inequality

$$\frac{1}{m-t}(y_2 - S_2\hat{\eta}(S_1))'\Sigma^{-1}(S)(y_2 - S_2\hat{\eta}(S_1))/\hat{\sigma^2}(S) \leq F \tag{8}$$

where F is the $1-\alpha$ quantile of an F distribution with $m-t, n(m-k)$ degrees of freedom

$$\Sigma(S) = I + S_2(S_1'S_1)^{-1}S_2',$$

$$\hat{\sigma^2}(S) = \frac{1}{n(m-k)}[\sum_{i=1}^{n} | y^i - S\hat{\eta^i}(S) |^2] \tag{9}$$

$$\hat{\eta^i}(S) = (S'S)^{-1}S'y^i = S'y^i,$$

and

$$\hat{\eta}(S_1) = (S_1'S_1)^{-1}S_1'y_1.$$

Assuming ε and ε^i are i. i. d. $N(0, \sigma^2 I)$, it follows from the classical theory about linear models that the prediction set (8) has coverage probability exactly $1-\alpha$. Also, without normality assumption of ε^i's, the coverage probability is asymptotically $1-\alpha$.

Souders and Stenbakken of NIST in a series of papers have been working on how to reduce the number of tests performed on products such as analog-to-digital converters. Their approach using the notation of a purely empirical linear model is summarized below.

Estimation of S.
Similar to most statistical approaches, Souders and Stenbakken used basically the principal components of y^i to estimate the column space of S. Precisely, let

$$Y = (y^1, \dots, y^n)$$

be the $m \times n$ matrix of the training data and perform a singular value decomposition on $\frac{1}{\sqrt{n}}Y$ so that

$$\frac{1}{\sqrt{n}}Y = (\widehat{S}, \widehat{S}_0)(\widehat{D})\widehat{V} \tag{10}$$

where $(\widehat{S}, \widehat{S}_0)$ and $\widehat{V}$ are orthogonal matrices of sizes m and n; $\widehat{S}$ is $m \times k$; and $\widehat{D}$ is an $m \times n$ diagonal matrix whose diagonal elements $\widehat{d}_i$'s, called the singular values are decreasing in i. In doing this decomposition, we follow the convention in (3). The decomposition is always possible for any matrix. Assuming that k is known, they then propose to estimate S by $\widehat{S}$, which has size $m \times k$ as well, and η^i by the estimator

$$\widehat{\eta}^i(\widehat{S}) = (\widehat{S}'\widehat{S})^{-1}\widehat{S}'y^i = \widehat{S}'y^i, \tag{11}$$

where for any matrix M, M' denotes the transpose of M.

Choice of t.
How does one choose t? A rule of thumb suggested by Souders and Stenbakken is that t should be four times as large as that of $\widehat{k}$, the estimate of k. Our statistical intervals, however, provide a more precise criterion: choose t big enough so that the intervals are narrow enough.

One new issue that arises in the prediction context is which subvector y_1, whose elements are called the points, should be chosen. This involves design of experiments. Additionally the dimensionality k is typically unknown and a question is how to estimate it. For the latter problem, there are suggested estimators in Yochmowitz and Cornell (1978), as well as Engler (1996). We shall address the problem about

Choice of Test Points.
The problem of choosing the test points is important, because a better set of test points gives shorter confidence intervals and prediction intervals. The algorithm described below is suggested by Souders and Stenbakken (see Koffman, Souders and Stenbakken (1994)) and seems to work well. This algorithm will be used in constructing our intervals and our numerical studies suggest that the resultant intervals have good coverage probabilities.

To illustrate, we shall assume that the estimated design matrix $\widehat{S}$ has size 8192×17. The problem of selecting say $t = 64$ points is equivalent to that of choosing 64 rows of $\widehat{S}$. Whether some of these 64 rows can be the same depends on the application. In Souders and Stenbakken's application, this is allowed. (If the rows should be all different, the following algorithm can be modified by imposing the constraint.) The algorithm of Souders and Stenbakken involves two stages. In the first stage, they choose the first 17 rows by applying a greedy algorithm. Precisely, the first chosen row of the algorithm has the largest norm. To choose the second row, they project all the other rows onto the first row and pick the row with the larger orthogonal distance from the first row. Similarly to include the next row, they project the rest of the rows onto the space spanned by the chosen rows and choose to include the row with the largest perpendicular distance from the space. This process terminates when 17 rows are chosen. This process can be carried out by applying a QR decomposition to $\widehat{S}'$. This first stage was also suggested by Gail and Kiefer (1980) and have been shown to be quite desirable in Johnson and Nachtshein (1983).

In the second stage, more rows are included one at a time. Each new row corresponds to the coordinate at y that has the larger prediction variance. To be precise, the prediction variances referred to are the diagonal elements of the matrix

$$\widehat{S}(\widehat{S}_1'\widehat{S}_1)^{-1}\widehat{S}'$$

where $\widehat{S}_1$ consists of the rows that have been chosen.

Point Predictor of y_2.
With $\widehat{S}$ estimated and y_1 observed, y_2 is predicted by $\widehat{S}_2\widehat{\eta}(\widehat{S}_1)$ where

$$\widehat{\eta}(\widehat{S}_1) = (\widehat{S}_1'\widehat{S}_1)^{-1}\widehat{S}_1'y_1.$$

Souders and Stenbakken and other users of their approach are very concerned about how to assess the accuracy of the prediction. We propose in this paper statistical intervals (or sets) whose asymptotic coverage probability can be shown to be equal to the targeted value $1 - \alpha$. This statistical interval or set can therefore be used to assess the error of prediction, to assure quality in the production, and to choose t as has been pointed out in a previous paragraph. Numerical studies show that the asymptotic prediction interval works well even when n is moderate.

3 Purely Empirical Linear Model

In this section, we shall focus on model (6), where k is assumed to be known. It is assumed that ε and $\varepsilon^i, i = 1, 2, \ldots, n$ are $(n+1)$ independent and identically distributed (i.i.d.) random vectors with a common zero mean vector and a common covariance matrix $\sigma^2\Sigma$ where σ^2 is unknown.

The matrix S is then estimated by $\widehat{S}$ in (10) and η^i by the least squares estimator $\widehat{\eta}^i(\widehat{S})$ in (11). Let $\hat{d}_i$ and d_i, respectively, denote the singular values of $Y/\sqrt{n}$ and S. Under the assumption that η^i are i. i. d. with identity second moment, it can be shown that

$$\hat{d}_i^2 \to d_i^2 + \sigma^2 \text{ for } i = 1, \ldots, k \tag{12}$$

and

$$\hat{d}_i^2 \to d_i^2 \text{ for } i > k$$

as $n \to \infty$. Furthermore $\widehat{S}$ is consistent, i.e.,

$$\hat{S} \to S. \tag{13}$$

See Hwang and Liu (1995a and b) for the proof of (12) and (13). The result provides us consistent estimators for S. Also (12) is obviously useful for estimating k when k is unknown. This will be discussed in Section 4.

Since S is unknown, it then seems reasonable to estimate S, S_1 and S_2 in (8) by their consistent estimates $\hat{S}, \hat{S}_1$ and $\hat{S}_2$ respectively. This leads to the prediction set for y_2:

$$\frac{1}{m-t}(y_2 - \hat{S}_2\hat{\eta}(\hat{S}_1))\Sigma^{-1}(\hat{S})(y_2 - \hat{S}_2\hat{\eta}(\hat{S}_1))/\hat{\sigma^2}(\hat{S}) \leq F_{1-\alpha}. \tag{14}$$

Here

$$\hat{\sigma^2}(\hat{S}) = (\sum_{i=1}^{n} | y^i - \hat{S}\hat{\eta^i} |^2)/[(n-k)(m-k)] \tag{15}$$

and $F_{1-\alpha}$ is the $1-\alpha$ upper quantile of the F distribution with $(m-t)$ and $(n-k)(m-k)$ degrees of freedom. See Hwang and Liu (1995a, b) for a detailed proof that (14) has asymptotic coverage probability $1-\alpha$.

Intervals for the components of y_2.
Obviously we could also construct prediction intervals for the jth component $y_{2(j)}$ of y_2. The point estimator of $y_{2(j)}$ is

$$\widehat{y}_{2(j)} = \widehat{S}'_{2(j)}\widehat{\eta}(\widehat{S}) \tag{16}$$

where $\widehat{S}'_{2(j)}$ is the jth row of $\hat{S}_2$. Then a prediction interval for $y_{2(j)}$ with a $1-\alpha$ asymptotic coverage probability is

$$|y_{2(j)} - \widehat{y}_2(j)|/[\widehat{\sigma}(\widehat{S})w_{jj}^{\frac{1}{2}}] \leq t_{1-\frac{\alpha}{2}} \tag{17}$$

where w_{jj} is the (j,j)th element of $\sum(\widehat{S})$ and $t_{1-\frac{\alpha}{2}}$ is the $1-\frac{\alpha}{2}$ quantile of the t-distribution with $(n-k)(m-k)$ degrees of freedom.

Scheffé's type simultaneous confidence intervals.
In Souders and Stenbakken's application and perhaps many other applications, it is important to construct confidence intervals for the expectation of y, which equals $S\eta$. The expectation of y will be denoted by $y^T = S\eta$, because this vector is considered to be the "truth" by the engineers.

As before, we shall first derive confidence sets for y^T, assuming that S were known. In fact we deal with η first. The standard asymptotic confidence set for η with $1-\alpha$ coverage probability is

$$(\eta - \widehat{\eta}(S))'(S_1'S_1)(\eta - \widehat{\eta}(S))/[k(\widehat{\sigma}^2(S)] \leq F. \tag{18}$$

The last statement holds whether η is fixed or random. Therefore estimating S leads to an asymptotically valid $1-\alpha$ confidence set

$$(\eta - \widehat{\eta}(\widehat{S}))'(\widehat{S}_1'\widehat{S}_1)(\eta - \widehat{\eta}(\widehat{S})) \leq kF_{1-\alpha}\widehat{\sigma}^2(\widehat{S}) \tag{19}$$

where $\widehat{\sigma}(\widehat{S})$ is defined in (15) and $F_{1-\alpha}$ is the $1-\alpha$ quantile of an F distribution with k and $(n-k)(m-k)$ degrees of freedom. The above inequality, by Scheffé type argument is equivalent to

$$|a'\eta - a'\widehat{\eta}|^2 \leq k(a'(\widehat{S}_1'\widehat{S}_1)^{-1}a)\widehat{\sigma}^2(\widehat{S})F_{1-\alpha}, \tag{20}$$

for every k-dimensional vector a, provided that $\widehat{S}_1'\widehat{S}_1$ is nonsingular. Hence the asymptotic probability that (20) holds for every a is $1-\alpha$. To derive a simultaneous Scheffé type interval for the jth component y_j^T of y^T, substitute $\widehat{S}_{(j)}'$, the jth row of $\widehat{S}$, for a'. This gives

$$|\widehat{S}_{(j)}'\eta - \widehat{S}_{(j)}'\widehat{\eta}|^2 \leq B_j^2 \tag{21}$$

where

$$B_j^2 = k\widehat{S}_{(j)}'(\widehat{S}_1'\widehat{S}_1)^{-1}\widehat{S}_{(j)}\widehat{\sigma}^2(\widehat{S})F_{1-\alpha}.$$

The probability that the above inequality holds for every j is at least $1-\alpha$ asymptotically. This probability is expected to be close to $1-\alpha$ if m is large. Replacing $\widehat{S}_{(j)}'\eta$ by $y_j^T = S_{(j)}'\eta$ leads to the inequality

$$|y_j^T - \widehat{S}_{(j)}'\widehat{\eta}|^2 \leq B_j^2. \tag{22}$$

The replacement will not change the probability content since m is fixed and hence

$$|S_{(j)}'\eta - \widehat{S}_{(j)}'\eta| \to 0$$

with probability one. The argument leads to the following theorem.

Theorem 1 *Assume that $S_1'S_1$ is nonsingular and ε has $N(0,\sigma^2 I)$ distribution. Then the event that inequality (22) simultaneously holds for every j, $1 \leq j \leq m$, has at least $1-\alpha$ asymptotic coverage probability.*

Inequality (22) obviously is equivalent to the assertion that y_j^T is in the interval

$$(\widehat{S}'_{(j)}\widehat{\eta} - B_j, \widehat{S}_{(j)}\widehat{\eta} + B_j). \qquad (23)$$

We could also derive a $1-\alpha$ non-simultaneous confidence interval for y_j^T for a fixed j. The interval is then (22) except that in the definition of B_j^2, k is replaced by 1 and $F_{1-\alpha}$ denotes the $1-\alpha$ quantile of an F distribution with 1 and $(n-k)(m-k)$ degrees of freedom.

Koffman and Stott (1993) applied HELP with a purely empirical linear model to calibrate multirange instruments that are machines measuring voltages of signals with wide variation in frequency, voltage, current, etc. To obtain exhaustive accurate NIST measurements of a multirange instrument costs millions of dollars and may take several months. There are exhaustively measured data available from the past that correspond to other similar instruments. Using the earlier notations, y^i, $1 \leq i \leq n = 100$, stands for the differences between the measurements of the ith multirange instrument and its accurate NIST measurements of m $(= 255)$ levels. By these training data, the dimension of the subspace k is estimated to be 20. With 50 chosen test points, they predict the rest of the points.

4 Partially Empirical Linear Model

Parallel to Section 3, statistical intervals and their properties can similarly be developed for the partially empirical linear model. See Hwang and Liu (1995a and b) for the details. In this section, we describe our motiving example where a partially empirical linear model is applied, and report some numerical study results.

Now we discuss our motivating example of Souders and Stenbakken (1985, 1987, 1990) and Stenbakken and Souders (1990, 1992). They are interested in testing whether a newly made device such as an analog-to-digital converter (A/D Converter) works properly. For a 13 bit A/D converter, to be absolutely sure, one needs to test 8192 outputs, corresponding to transition levels (usually voltage levels) for the conversion of the analog signals to the digital signals. The behavior of the converter is largely determined by a relatively small number of coefficients (less than 100) such as resistance, capacitances or transistor transconductances. A natural question is whether one can choose to test at only a small number of levels and still provide a reliable conclusion.

Using the notation in Section 1, y denotes the vector of error transition levels, i.e., the 8192 dimensional vector whose components are the difference between the observed transition levels and the targeted transition levels. To study their problem, we apply a partially empirical linear model (4) to y. The known design matrix X is of size 8192×14, which is the design matrix corresponding to the 2^{13} factorial design (see, for example, Table 12.1a on p. 377 of Box, Hunter and

Hunter (1978)) and the last column consists of ones. By using the exhaustive measurements of the $n = 88$ converters and y_1 the subvector of only $t = 64$ coordinates of y, HELP predicts well the behavior of y_2 the remaining 8128 transition levels. This is probably due to the fact that the means of y and y^i sit on a linear subspace with low dimensionality $k + l$ whose estimated value is about 17. In manufacturing converters, testing cost constitutes 20 to 50% of the total manufacturing cost. Obvious, a saving from 8192 measurements down to 64 measurements, less than one percent, can reduce production cost tremendously.

In the numerical study, we "experiment" with the data to check the coverage probability of the prediction set of y_2. The data we have are 88 vectors. Each vector consists of 8192 error transition levels of a 13–bit converter. Additionally we have three repeated exhaustive measurements of converter. We divide each of the 88 vectors into 256 "experiments". In each "experiment" there are 32 error transition levels. The ith data in the jth experiment is denoted by y_j^i, $1 \le i \le 88$ and $1 \le j \le 256$. Hence for each experiment $n = 88$ and $m = 32$.

In reality, each y_j^i represents the error transition levels of a 5–bit converter. In each experiment, we use $t = 25$ testing points to predict the rest of the 7 points of a chosen "future" vector based on the other 87 "training data". The chosen "future" vector, in the first experiment, is the first vector. Therefore, the training data are $y_1^2, \ldots, y_1^{88}$ which are used with the first 25 elements of y_1^1 to "predict" the rest of the 7 elements of y_1^1. In the second experiment, the "future" observation corresponds to the second converter y_2^2. Similarly, in the 88th experiment, the "future" observation corresponds to the 88th converter y_{88}^{88}. For the 89th experiment, the future observation corresponds to the first converter y_{89}^1. Continuing this way, we could apply each of our procedures 256 times and calculate the relative frequency that the prediction interval covers the target. The relative frequencies given below in Table 1 are then used to estimate the coverage probabilities.

Table 1. Relative frequencies of prediction sets based on various approaches. The nominal level is .95.

		Residual		Repeated		Pooled	
		No Select	Select	No Select	Select	No Select	Select
physical	$\widehat{k}_L$	.695	1.00	.098	.293	.691	1.00
	$\widehat{k}_A$	.781	1.00	.098	.293	.777	1.00
purely	$\widehat{k}_L$	.797	.828	.984	.992	.820	.883
empirical	$\widehat{k}_A$	.855	.918	.957	.996	.859	.922
partially	$\widehat{k}_L$	.883	.887	.984	.992	.898	.926
empirical	$\widehat{k}_A$	.914	.973	.980	.996	.918	.949

In the table, "physical" refers to a physical model only, i.e., model (1) without μ and χ. Here the first five columns of X is the design matrix corresponding to the 2^5 factorial design and the last column consists of ones.

"Residual", "Repeated" and "Pooled" in Table 1 refer to the procedures where σ^2 is estimated respectively by (15), the sample variance of the repeated measurements, and the pooled estimator of (15) and Repeated. "Select" and "No Select" refer to whether one uses the selection method described in Section 2 to choose the test points. In the "No Select" column, we always use the first 25 points y_1 to predict the rest of the 7 points y_2. Finally the estimators of k are

$$\widehat{k}_L = \text{ the minimum of } t-\ell \text{ and the largest } i's \text{ such that } \widehat{d_i^2} > \sigma^2 \tag{24}$$

and

$$\widehat{k}_A = \text{ the minimum of } t-\ell \text{ and the largest i's such that the average of } \widehat{d_i^2}, \widehat{d_{i+1}^2}, \ldots \widehat{d_{m-\ell}^2} \text{ is greater than } \sigma^2. \tag{25}$$

Conclusion

In a traditional multivariate linear model, the design matrix is prespecified. In this paper, we start with such a model. We allow the flexibility of adding more empirical columns which are estimated by the data. The number of columns added is also determined by the data.

The results of using empirical columns are quite remarkable. The numerical studies clearly show that there are advantages of introducing these empirical columns in terms of the coverage probabilities.

We show how to apply the technique to two engineering problems to save measurements. The point estimation procedure to save measurement was developed by Souders and Stenbakken. The paper developed statistical intervals for quantifying the uncertainty.

References

[1] Anderson, T. W. (1984), *An introduction to multivariate statistical analysis*, Second ed., Wiley Series in Probability and Mathematical Statistics.

[2] Azzalini, A. and Diggle, P. J. (1994), *Prediction of soil respiration rates from temperature, moisture, moisture content and soil type*, Applied Statistics 43, 505–526.

[3] Box, George E. P., Hunter, W. G., and Hunter, J. S. (1978), *Statistics for Experimenters*, John Wiley & Sons, Inc., New York.

[4] Chung. K. L. (1974), *A Course in Probability Theory*, Academic Press, New York and London.

[5] Engler, H. (1996), *Determining model dimension with sphericity tests*, Tech. Rept. of Electro System Group at NIST.

[6] Gabriel, R. R. (1978), *Least squares approximation of matrices by additive and multiplicative models*, JRSS B 40, 2, 186-196.

[7] Galil, Z. and Kiefer, J. (1980), *Time– and space–saving computer methods, related to Mitchell's DETMAX, for finding D–optimum*, Technometrics 22,3, 301–313.

[8] Hwang, J. T. and Liu, H. K. (1995a), *High–dimensional empirical linear prediction with application to quality assurance in industrial manufacturing*, Cornell Tech. Rept.

[9] Hwang, J. T. and Liu, H. K. (1995b), *Asymptotic theory for prediction through empirical linear models*, Spring Research Conference: 1994 Proceedings, 207–216.

[10] Hwang, J. T. and Liu, H. K. (1995c), *Applications of empirical linear prediction*, Physical and Engineering Sciences: Proceedings of the 1994 Joint Statistical Meetings, 175–180.

[11] Johnson, M. E. and Nachtsheimn, C. J. (1983), *Some guidelines for constructing exact D–optimal designs in convex design spaces*, Technometrics 25, 271–277.

[12] Koffman, A. D. and Stott, H. C. (1993), *Modeling and test point selection for a thermal transfer standard*, Proceedings of the NCSI Workshop and Symposium, Albuquerque, NM, 199–310.

[13] Okamoto, M. (1973), *Distinctness of the eigenvalues of a quadratic form in a multivariate sample*, Annals of Statistics 1, 763-765.

[14] Scheffé, H. (1959), *The analysis of variance*, John Wiley & Sons, Inc., New York.

[15] Souders, T. M. and Stenbakken, G. N. (1985), *Modeling and test point selection for data converter testing*, IEEE International Test Conference Proceedings, IEEE Press, New York, NY, 813-817.

[16] Souders, T. M. and Stenbakken, G. N. (1987), *Test point selection and testability measures via QR factorization of linear models*, IEEE Trans. Instrum. Meas. Vol. IM-36, No. 2, 406-410, June.

[17] Souders, T. M. and Stenbakken, G. N. (1990), *A comprehensive approach for modeling and testing analog and mixed-signal devices*, 1990 International Test Conference, IEEE Computer Society Press, Sept.

[18] Stenbakken, G. N. and Souders, T. M. (1990), *Linear error modeling of analog and mixed-signal devices*, Proceedings, 1991 International Test Conference, IEEE Computer Society Press, Sept.

[19] Stenbakken, G. N. and Souders, T. M. (1992), *Developing linear error models for analog devices*, Tech. Rept. at NIST.

[20] Tyler, D. E. (1981), *Asymptotic inference for eigenvectors*, Annals of Statistics 9, 4, 725-736.

[21] Yochmowitz, M. D. and Cornell, R. G. (1978), *Stepwise tests for multiplicative components of interaction*, Technometrics 20, 79–84.

Advanced Mathematical Tools in Metrology III
Edited by P. Ciarlini, M.G. Cox, F. Pavese & D. Richter

WAVELET METHODS IN SIGNAL PROCESSING

P. MAASS

Universität Potsdam, Institut für Mathematik, Am Neuen Palais 10, D-14415 Potsdam, Germany

This paper aims at developing a guideline on how to select an appropriate wavelet and how to interpretate the resulting wavelet transform for specific applications. The diversity of wavelet methods requires a detailed mathematical analysis of the underlying physical or technical problem in order to take full advantage of this new toolbox.

Data sets from real-life applications are used to demonstrate the scope of wavelet methods in signal processing.

1 Introduction

During the last decade wavelet methods have developed into powerful tools for a wide range of applications in signal and image processing. The diversity of wavelet methods, however, requires a detailed mathematical analysis of the underlying physical or technical problem in order to take full advantage of this new toolbox. This paper aims at developing a guideline on how to select an appropriate wavelet and how to interpretate the resulting wavelet transform data for specific applications.

We start with an introduction on the basic properties of the continuous wavelet transform followed by an explicit example. The subsequent chapter deals with the discrete wavelet transform and its application to the evaluation of ECG-data. We particularly emphasise the importance of choosing an adequate wavelet for each task.

2 Continuous wavelet transform

The continuous wavelet transform was introduced in order to overcome the limited time-frequency localization of Fourier methods for non-stationary signals. Many papers have been written on the differences and similarities between Fourier and wavelet transforms, for more detailed information on this topic see e.g. [5,4].

The wavelet transform correlates the signal f with a shifted and translated test function ψ, $(a, b \in \mathbb{R}, a \neq 0)$:

$$L_\psi f(a,b) = |a|^{-1/2} \int\limits_{I\!R} f(t)\psi\left(\frac{t-b}{a}\right) dt \quad . \tag{1}$$

The parameter a determines the scale (or size of details) which is examined, the scale becomes finer and finer as a approaches 0. This property has lead to the interpretation of the wavelet transform as a mathematical microscope.

A pair of functions $(\psi, \tilde{\psi})$ is called a wavelet pair, if it satifies the admissibility condition, ($\hat{\psi}$ denotes the Fourier transform of ψ):

$$0 < \int\limits_{I\!R} \frac{|\hat{\psi}(\omega)\widehat{\tilde{\psi}}(\omega)|}{|\omega|} d\omega < \infty \quad . \tag{2}$$

This condition insures that f can be recovered from its wavelet transform $L_\psi f$ by

$$f(t) = \int\limits_{I\!R}\int\limits_{I\!R} L_\psi f(a,b)\ |a|^{-1/2}\ \tilde{\psi}\left(\frac{t-b}{a}\right) \frac{dadb}{a^2} \quad . \tag{3}$$

Some of the most interesting cases arise when $\psi = \tilde{\psi}$, in which case ψ is called a wavelet. The admissibility condition then essentially states that ψ has zero mean:

$$\int\limits_{I\!R} \psi(t)\ dt = 0 \quad . \tag{4}$$

Standard examples for wavelets are:

- Haar wavelet: $$\psi(t) = \begin{cases} 0 & \text{for } t < 0 \text{ or } t > 1 \\ 1 & \text{for } 0 \le t < 1/2 \\ -1 & \text{for } 1/2 \le t \le 1 \end{cases} \tag{5}$$
- Mexican hat wavelet: $$\psi(t) = -\frac{d^2}{dt^2}e^{-t^2/2} = (1-t^2)e^{-t^2/2} \ , \tag{6}$$
- Morlet wavelet: $$\psi(t) = \pi^{-1/4}\left(e^{-i\ell t} - e^{-\ell^2/2}\right) e^{-t^2/2} \ , \tag{7}$$

  the optimal choice of the parameter ℓ depends on the application.

Figure 1 depicts the wavelet transform of a chirp–signal

$$f(t) = sin\left(t^{5/2}\right) \quad . \tag{8}$$

This is a typical signal which is difficult to analyse with Fourier or windowed Fourier techniques since its local frequency increases polynomially, or equivalently the size of details decreases. It therefore exhibits a truly multiscale

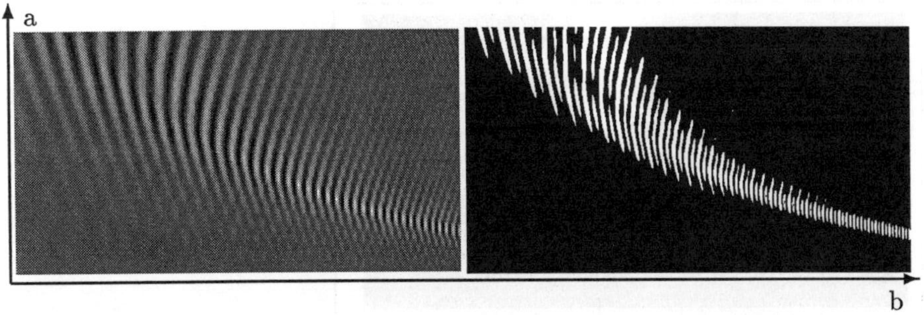

Figure 1: The mexican hat was used to compute the wavelet transform of the chirp signal. The plot on the right shows the extrema of the wavelet transform, they show that the size of details decreases continuously.

structure. Figure 1 also displays its wavelet transform in the upper (b, a)–plane after thresholding, the local extrema clearly reveal the decreasing size of details.

Most applications in signal processing require more than a visual inspection of the wavelet transform. Therefore we list some of the most important tools for interpretating the wavelet transform. We always start from a mathematical property of $L_\psi f(a, b)$ followed by a description of its impact in signal processing.

2.1 Asymptotic behaviour for $a \to 0$

Let us consider the Haar wavelet and its wavelet transform. $L_\psi f(a, b)$ then denotes the (weighted) difference between the average of f over the interval $[b, b + a/2)$ and the average of f over the interval $[b + a/2, b + a)$. As $a \to 0$ this approximates the first derivative of f at $t = b$.

More general one can define the "order of a wavelet", [11], depending on the asymptotic behaviour of $L_\psi f(a, b)$ as $a \to 0$. E.g. the Haar wavelet is of order 1, the mexican hat wavelet is of order 2.

Lemma *If ψ is a wavelet of order k and if f is k–times differentiable at $t = b$, then*

$$|a|^{-k-1/2} \; L_\psi f(a, b) \longrightarrow c \; f^{(k)}(b) \quad as \quad a \to 0 \; . \tag{9}$$

The asymptotic behaviour of the wavelet transform therefore allows a local regularity estimate for f. This pointwise regularity analysis could not be carried out with the Fourier transform, since its asymptotic behaviour as $\omega \to \infty$ only

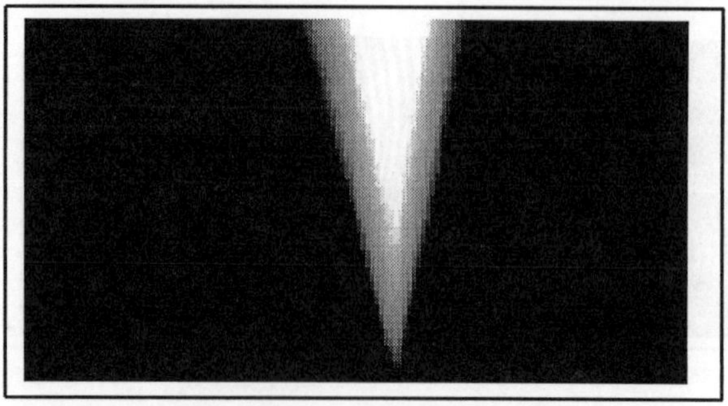

Figure 2: Choosing the Haar wavelet and displaying its wavelet transform reveals discontinuities of the signal. In this case a smooth signal with a small defect at $t = 0.2$ was analyzed. The plot displays the wavelet transform for $b \in [-1, 1], a \in [0, 0.5]$.

yields a global regularity estimate.

In signal processing this property can be used to detect and localize discontinuities (jumps, cracks, defects): $L_\psi f(a, b)$ decays to zero at regular points $t = b$ but diverges at points where f is not sufficiently differentiable.

More precisely: depending on the mathematical model for the discontinuity one may choose an appropriate wavelet. If e.g. one searches for hidden jumps – they relate to points where f is not differentiable – then an analysis with the Haar wavelet is sufficient to detect those irregularities.

Figure 2 displays the wavelet transform of a smooth function $f(t) = e^{-t^2/2}$ with a small defect at time $t = 0.2$. The Haar wavelet was used in this example.

2.2 Wavelet extrema

The most obvious interpretation of the wavelet transform is based on the local extrema of $L_\psi f(a, b)$.

Let us consider the problem of detecting a pattern whose shape is similar to ψ but whose size is unknown. In this case a pattern matching approach with varying pattern size is needed.

The wavelet transform does exactly that: it correlates the signal f with shifted and scaled version of the pattern ψ, hence a local extrema of $L_\psi f(a, b)$ reveals a searched for pattern of size a at time $t = b$.

Note that – since wavelets have zero mean – this analysis is independent of shifts by constant. Moreover if ψ has more vanishing moments then the computation is independent of linear or polynomial trends in the data f.

2.3 Indicator function

Suppose that the scale (size) of a searched for pattern is unknown in some application. The scale – energy spectrum or indicator function

$$n(a) = \int_{\mathbb{R}} |L_\psi f(a,b)|^2 \, db \tag{10}$$

sums the L^2–energy of all details on a specified scale. If the energy is evenly distributed over all scales, then $n(a)$ is exptected to exhibit a regular decay as $a \to 0$. In any case a deviation or local maxima of $n(a)$ reveals that f has some significant structure on the scale a.

For some classes of functions more specific power laws can be proved. E.g. turbulent signals have decay rates

$$n(a) \sim a^\kappa, \tag{11}$$

where κ varies between 5/3 (fully developed turbulence) and 3 (non turbulent flows). This can be used to classify turbulence measurements, see e.g. [10], where wind measurements were evaluated in order to detect cirrus clouds. Similar experiments are presently carried out by D. Handorf, Deutscher Wetterdienst, Potsdam, who is examining antartical wind measurements.

2.4 Fractal and multifractal structures

Fractal structures are natural multiscale objects. They are therefore excellent candidates to demonstrate the usefulness of wavelet techniques.

The above defined indicator function can be refined by determining the decay rates for different energy measures:

$$n(a,q) = \int_{\mathbb{R}} |L_\psi f(a,b)|^q \, db \, . \tag{12}$$

The asymptotic behaviour as $a \to 0$

$$n(a,q) \sim a^{D(q)} \tag{13}$$

can be used detect and classify fractal and multifractal structures. For detailed examples and more information how to interpret $D(q)$ see[1].

2.5 Wavelets and bandpass filtering

For a fixed value of a one can view the wavelet transform as a convolution of f with a scaled wavelet ψ. The convolution property of the Fourier transform gives the following result.

Lemma: *Assume that* $\psi,\ f \in L^2(\mathbb{R})$. *The one–dimensional Fourier transform of* $L_\psi f(a,b)$ *with respect to the variable* b *yields*

$$\widehat{L_\psi f}(a,\omega) \ = \ c|a|^{1/2}\hat{f}(\omega)\hat{\psi}(a\omega) \ . \qquad (14)$$

For fixed a the wavelet transform of f can thus be viewed as a bandpass filter with filter function $\psi(t/a)$. Since the admissibility condition forces the Fourier transform of ψ to vanish at $\omega = 0$ this is indeed a bandpass filter with a variing frequency band depending on a. The wavelet transform therefore simultaneously computes bandpass filters for a complete family of frequency bands. For a more detailed description see [4].

3 Detection of defects

We now apply some of the above described tools for solving a problem in signal processing. Check–up procedures for industrial machinery, which require to stop the production for examination, lead to a slow down in the overall production process. Neglecting regular check–ups however will lead to unexpected breakdowns which result in even more costly interruptions of the production.

Sensor techniques can offer an alternative in certain enviroments. Figure 3 displays measurements taken with a vibration sensor, which was placed on a machine in textil production. The data shows a low frequency periodic background signal perturbed by measurement noise.

The problem is to detect any potential defects (cracks, dirt, etc.) which show as additional structures in the noisy signal. Defects are modelled as follows in this setup: a defect starts at time t_0 and produces an error component of the type

$$ae^{b(t-t_0)}cos(k(t-t_0)) \quad \text{for} \quad t > t_0 \ . \qquad (15)$$

The problem is to detect such defects in the noisy data and to determine the parameters (a, b, k, t_0) in order to classify the defect.

This problem has a multiscale structure since the underlying low frequency component, the measurement noise and the searched for pattern (defect) live on different time scales. Moreover the time scale of the defect is unknown.

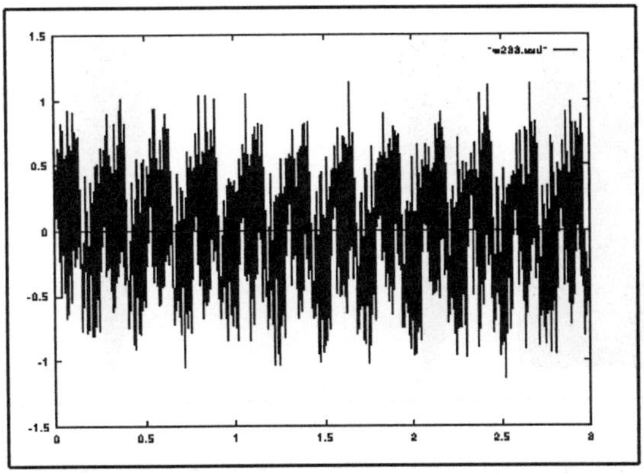

Figure 3: A data set obtained with a sensor which measures vibrations. The problem is to detect defects in the noisy signal. This set of data was provided by Prof. Dr. H. Dresig, University of Chemnitz.

Of course this problem can be tackled by standard techniques in signal processing (Fourier methods, bandpssfilter, pattern matching). But our aim is to demonstrate that wavelet techniques offer a potentially superior alternative, which – due to its multiscale structure – lead to a precise and stable analysis of this problem.

Before starting to analyze any signal with wavelet methods one should answer three basic questions:

- Why should wavelet analysis help to solve this specific problem?
- Which wavelet should be used?
- Which tools for interpretating the computed wavelet transform will reveal the desired information?

The first question can be answered positively whenever the signal has some multiscale structure or if the searched for information lives on an a priori unknown scale. Our problem of detecting defects in a noisy signal lives on three length scales: the slow frequency of the bakcground signal, the noise components and the search for defects, they comprise details of a yet unknown size. The wavelet transform allows, due to its bandpass filtering property, to scan the signal on different frequency bands or detail scales simultaneously.

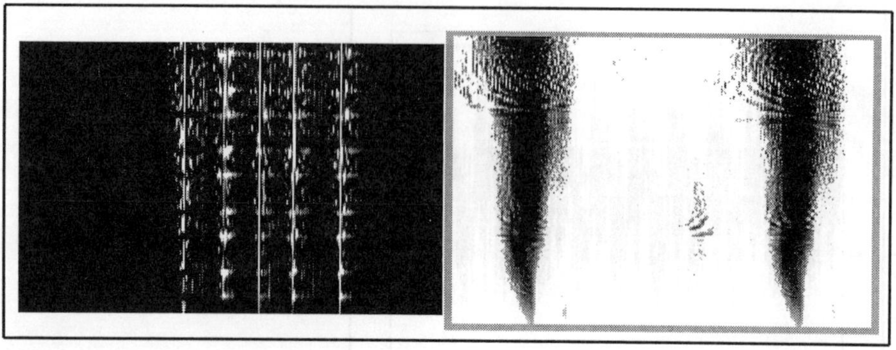

Figure 4: The Haar wavelet was used to detect the time instances of the defects. The plot on the left displays the wavelet transform after thresholding. The plot on the right shows a zoom into a segment containing two defects. The defects can be localize with high accuracy.

The scales between the noise and the background oszillation contain the desired information.

The answer to the second question requires to study the mathematical background of the problem at hand. Our test example mainly asks for two different types of information, namely to determine the time instances of the defects and to determine some parameters (frequency, amplitude, duration) of the defect. The model for the defects – they are assumed to start with a sharp jump – asks to search for hidden discontinuities. This can be done by using the Haar wavelet and displaying the thresholded wavelet transform, see Figure 4. The extrema of the wavelet transform clearly indicate critical segments of the signal.

Now we can examine those segements of the signal which contain the relevant information. Let us concentrate on determining the local frequency k of the defect. The Morlet wavelet typically gives good results when searching for local oscillations. Hence we perform a corresponding wavelet transform on those segments. The adequate tool for interpretating this transform is the indicator function $n(a)$. It shows a clear local maximum at some scale $a^* \sim 0.016$ which translates into the local frequency

$$k = \ell / a^* \quad , \tag{16}$$

where ℓ denotes the parameter of the Morlet wavelet. In our experiment we choose $\ell = 5$, moreover only the real part of the Morlet wavelet was used for the computations.

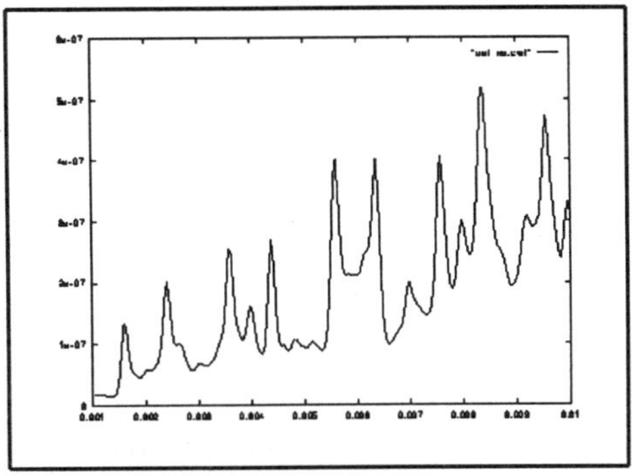

Figure 5: The smallest value of $a^* \sim 0.016$, where $n(a)$ shows a significant deviation from a regular decay, indicates the scale on which the defect lives. Extremal values for larger values of a are octave effects or perturbations on a larger scale.

4 Discrete wavelet transform

The continuous wavelet transform allows a precise and stable data analysis, however its evaluation is rather costly in terms of computer time. E.g. the wavelet transform depicted in Figure 1 was discretized on a 256×128 grid in the (b, a)–plane and the computation of $L_\psi f(a, b)$ at a single pair (b, a) requires the evaluation of an integral by a quadrature rule. We used Simpson's rule with approximately 100 discretization points.

Such a detailed analysis is not needed for many applications. The name discrete wavelet transform refers to the evaluation of the wavelet transform on a dyadic grid

$$d_k^m = L_\psi f(2^m, 2^m k) \ . \tag{17}$$

Note that the discretization of the shift variable b becomes finer and finer for small scales a, hence even local high frequency/small scale events can be detected reliably. All the properties and tools described in Chapter 2 can naturally be expressed in terms of the discrete wavelet transform $\{d_k^m\}$.

A major breakthrough for wavelet applications came with the discovery of Mallat's algorithm, which – for a special class of wavelets – allows an extremely fast evaluation of the discrete wavelet transform. This algorithm rests on a so called scaling function φ, which is assumed to obey a scaling equation

$$\varphi(t) = \sqrt{2} \sum_{\ell \in \mathbb{Z}} h_\ell \, \varphi(2t - \ell) \tag{18}$$

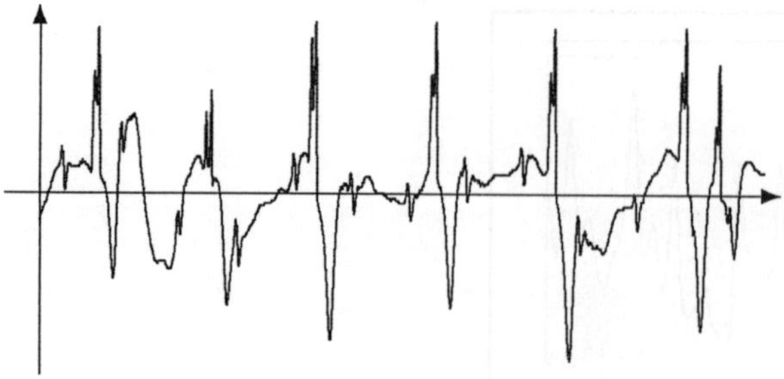

Figure 6: The aim is to locate the small structures stemming from the activities of the cardiac valves.

and which is related to the wavelet ψ via

$$\psi(t) \;=\; \sqrt{2} \sum_{\ell \in \mathbb{Z}} g_\ell \, \varphi(2t - \ell) \; . \tag{19}$$

The coefficient sets $\{h_\ell, g_\ell\}$ have a finite length in all practical cases. A simple example arises when $\varphi(t) = \chi_{[0,1]}(t)$, the characteristic function of the unit interval. This yields the Haar wavelet with $h_0 = h_1 = 1, g_0 = 1, g_1 = -1$. Let us now assume that the given data consists of a finite sequence

$$c_k^0 \;\; , \; k = 1, .., n \; , \tag{20}$$

of measured data. Via the scaling function one constructs a related data function

$$f(t) = \sum_{k \in \mathbb{Z}} c_k^0 \, \varphi(t - k) \;\; . \tag{21}$$

The computation of its discrete wavelet transform $d_k^m \;=\; L_\psi f(2^m, 2^m k)$ for $m > 0$ can now be carried out by recursive finite convolutions ($m > 0$):

$$d_k^m = \sum_{\ell \in \mathbb{Z}} g_{\ell - 2k} \; c_\ell^{m-1} \tag{22}$$

$$c_k^m = \sum_{\ell \in \mathbb{Z}} h_{\ell - 2k} \; c_\ell^{m-1} \;\; . \tag{23}$$

Due to the subsampling by a factor of 2 the length of the data sequence is approximately halfed in each recursion, this leads to a $O(n)$ complexitiy of the overall algorithm. This is faster than the fast Fourier transform!

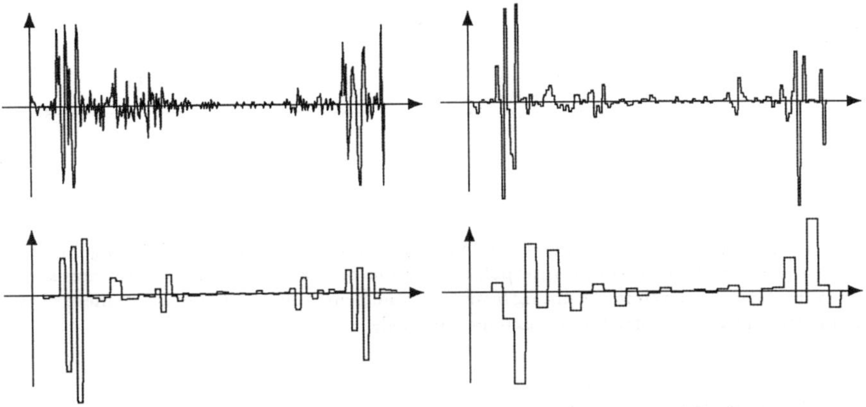

Figure 7: The discrete wavelet transform of the ECG–signal from a single heartbeat on scales $m = 1, .., 4$. The correspond to cuts through the continuous wavelet transform for $a = a_0\ 2^m$, where a_0 denotes the basic lenghtscale of the signal.

We will not describe the various families of discrete wavelets which have been constructed for a wide range of purposes. For a detailed and very readable introduction to the discrete wavelet transform see [12]. We end this section by describing an application using the discrete wavelet transform.

4.1 Analysing ECG signals

ECG–signals measure the electrical potential related to activities of the heart muscles. The large peaks in Figure 6 stem from the contractions of the large heart muscle, the smaller repeated pattern indicates the activities of the left and right cardiac valves. The detection of these patterns is vital for classifying arhythmia which may lead to heart failures.

The shape of this pattern does not change significantly even over several hours. However the width/size of this pattern differs according to the heart rate which may vary between 40 – 200. This data sequence was evaluated using a Daubechies wavelet with six coefficients $\{h_\ell, g_\ell\}$. This wavelet exhibits a shape similar to the valve pattern. We display the discrete wavelet transform on different scales, see Figure 7,

$$d^m = \{d_k^m\}\ , m = 1, 2, ... \tag{24}$$

they correspond to cuts through the continuous wavelet transform on different scales $a_m = 2^m a_0$, where a_0 is the basic stepwidth of the discretized signal.

After extracting the activity of the large heart muscle by standard techniques a simple thresholding locates the valve activities: they occurred on scale

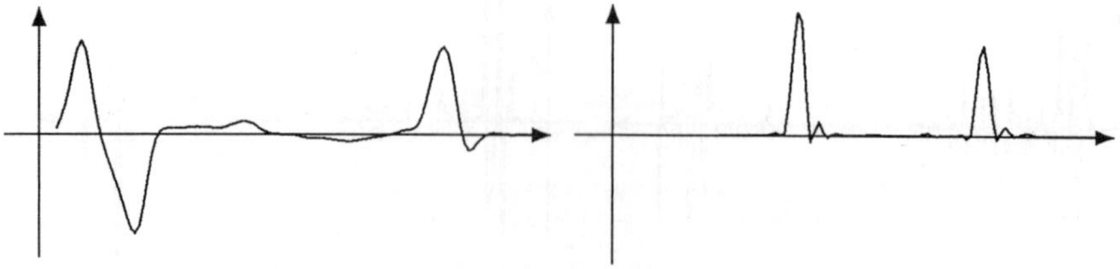

Figure 8: The plot on the left shows a smoothed version of the original ECG–signal. The plot on the right shows a spline–interpolated version of the d^3 discrete wavelet transform after extracting the maxima stemming from the large heart muscle. For enhancement of the maxime the data was plotted on an exponential scale.

$m = 3$ for this data segment.

Figure 8 displays a spline interpolation of scale $m = 3$ on an exponential scale, which emphasizes the local maxima.

5 Conclusions

This paper aimed at describing some basic properties of the wavelet transform which are relevant for applications in signal processing. In order to take full advantage of this toolbox a mathematical analysis of the underlying physical/technical problem is essential for choosing an appropriate wavelet and for interpretating the wavelet transform.

We have only sketched some wavelet techniques, for further and more refined methods the interested reader is referred to available textbooks, see e.g. [8,5,12]. Research articles on specific fields of applications are published e.g. in a wavelet series by Academic Press [3].

In particular we have not mentioned the application of wavelets in image processing, which is probably the most succesfull topic for wavelet analysis, see e.g. [9,2]. We have also neglected the applications of wavelet methods for solving operator equations and partial differential equations, see e.g. [7,6]. Finally we want to mention the "Wavelet Digest", the largest worldwide bulletin board for new wavelet developements, for subscritpion contact "wavelet@math.scarolina.edu".

Acknowledgements

The work of the author has been partially supported by the Deutsche Forschungsgemeinschaft under grant Ma 1657/1–2.

References

1. A. Arneodo, F. Argoul, J.F. Muzy, B. Pouligny, E. Freysz: *The optical wavelet transform*, in *Wavelets and their Applications*, eds. M.B. Ruskai, G. Beylkin, R. Coifman, I. Daubechies, S. Mallat, Y. Meyer, L. Raphael, Jones and Bartlett pubs., Cambridge MA, 1992
2. R. Bischoff, T. Boskamp, V. Dicken, P. Maaß, H. Peters, H.-G. Stark: *Bilddatenkompression mit Wavelet-Methoden*, Springer Sonderband, BMBF-Projekte *Anwendungsorientierte Mathematik*, Eds. K.-H. Hofmann, W. Jäger, Springer Verlag (1996)
3. C. Chui (Ed.), Series on *Wavelet Analysis and its Applications, vol. I - V*, Academic Press, New York, 1992-96
4. P. Kumar, E. Foufoula-Georgiou: *Wavelet analysis in geophysics: an introduction*, in *Wavelets in Geophysics*, eds. E. Foufoula-Georgiou, P. Kumar, Academic Press, 1994
5. Daubechies, I.: *Ten lectures on wavelets*, SIAM, Philadelphia (1992)
6. Dahmen, W., Prößdorf, S., Schneider, R.: *Wavelet approximation methods for pseudo-differential operators II*, Advances Comp. Math. **1** (1993) 259-335
7. Dicken, V., Maaß, P.: *Wavelet-Galerkin methods for ill-posed problems*, J. Inv. Ill-Posed Prob. (to appear 1996)
8. Louis, A. K., Maaß, P., Rieder, A.: *Wavelets - Theorie und Anwendungen*, Teubner, Stuttgart (1994). English version: Wiley, Chichester (to appear 1996)
9. Maass, P., Stark, H.-G.: *Wavelets and Digital Image Processing*, Surv. Math. Ind. 4, S.195-235 (1994)
10. M. Quante: *Some aspects of turbulence in the free atmosphere and its relation to cirrus clouds*, Proc. Fourth Workshop on ICE/EUCREX, (1991)
11. Rieder A.: *Wavelet Transform on Sobolev Spaces and its Approximation Properties*, Numer. Math., (1992)
12. M. Vetterli, J. Kovacevic: *Wavelets and Subband coding*, Prentice Hall PTR, New Jersey, 1995

Advanced Mathematical Tools in Metrology III
Edited by P. Ciarlini, M.G. Cox, F. Pavese & D. Richter

SOFTWARE PROBLEMS IN CALIBRATION SERVICES - A CASE STUDY

NORBERT GREIF, HEIKE SCHREPF and DIETER RICHTER
Physikalisch-Technische Bundesanstalt
Abbestr. 2-12, 10587 Berlin, Germany
email: dieter.richter@ptb.de

An approach is presented how to deal with software problems in calibration by illustrating a particular case. The case concerns the gauge block calibrations in length measurement. Risks for incorrect calibrations due to the software used are identified. The risks are classified as questions of completeness, correctness and adequacy of software with regard to the contents, problems arising from the incorrectness of the code, problems with data handling, problems of improper use of software, and management problems.

General recommendations from the viewpoint of software technology are given. To exclude, or at least minimize, risks of software failures, approaches are briefly outlined how to implement the results of the analysis into the practical work of calibration services considering the experience of software technology.

1 Introduction

Parts of measurement functions of calibration procedures have gradually become supported by software or have been even replaced by it. Without reducing the accuracy requirements, additional kind of problems has come up in metrology in general and calibration in particular: errors of measurement caused by software failures. As a consequence, this type of problems has to be dealt with in addition to the problems of precise measurements encountered anyway.

A variety of methods to back up software quality has been developed on the basis of advanced software technology. The definition of requirements plays a central role in the application of such methods. However, requirements for measurement systems are usually available for the system as a whole but not explicitly for its software part. This gap is of a basic nature. In our opinion, this is the reason why a real synergy between the metrology on the one side and the software technology on the other side does not yet exist. As it can be seen so far, approaches to bridge this gap are either one-sided or too general (for instance [5]).

In the field of length measurement, parallel gauge blocks are material measures which can be produced with the smallest uncertainty of measurement. They are, therefore, the most important material measures in length measurement. The parallel gauge block represents a certain length by the distance between two parallel boundary planes. Figure 1 shows the principle of distance measurement by the example of a comparison between a gauge block under test and a reference standard.

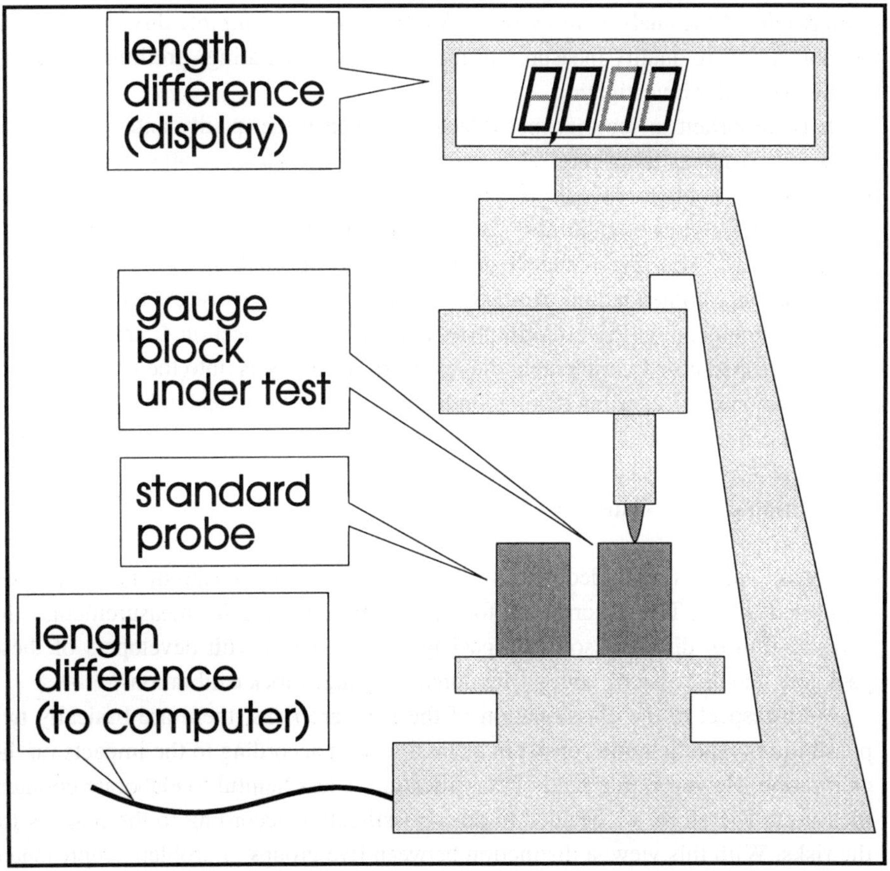

Figure 1: Principle of the difference measurement

The requirements for the measurement conditions are very high, in particular concerning the temperature and the care with which the blocks must be handled.

Suggested by the German Calibration Service DKD, Technical Committee for Length Measurement, it has been decided to carry out a study aimed at a detailed analysis of impacts of software failures on the reliability of calibrations. The investigations are restricted to gauge block calibrations based on mechanical difference measurements of parallel gauge blocks. On the basis of this analysis, suitable measures should be proposed to reduce the risks of software failures.

The aim of the paper is to present the approach which has been chosen together with results of the analysis in extracts. Another aim is to promote discussions about suitable software quality measures in metrology by illustrating preliminary results of this particular investigation.

It is important to see that many factors influence the quality and reliability of software. Among them are also such factors which are often regarded as unimportant. Though it would be possible to generalize some of the results in the regard to other types of calibration, we refrain from doing this in this paper.

In section 2, we give a survey of the results of the risk analysis. In section 3, theoretical recommendations from the point of view of software technology are given to exclude, or at least to minimize, risks. Finally, in section 4, approaches are briefly outlined how to implement the results of the analysis into the practical work of calibration services using recommendations from software technology.

2 Additional Risks Caused by Software

The risk analysis included the study of the length measurement principles, supported by PTB laboratory for interferential length measurement, the investigation of different software packages, discussions with developers of these packages and discussions with professionals of gauge block calibration.

With respect to the classification of the software risks, there are basically two possibilities. The first one consist in a classification according to the impacts on the calibration. However, this kind of classification is less helpful to elaborate counter-measures. Therefore, we decided to do a classification according to the reasons for the risks. With this view, a distinction between five groups of problems represented by the following subheadings seems to be useful.

2.1 Completeness, correctness and adequacy with regard to the contents

In general, this group is a very important one. All the problems concerning the adequate and correct coverage of expected functions by the software from the contents point of view belong to this class. In particular, questions arising from the application of mathematical algorithms, including the mathematical background and numerical aspects, play an essential role. However, in gauge block calibration no mathematical problems are actually encountered.

The functions supported by software are the control of gauge block comparisons and some aspects of measurement data processing.

The controlling part has to ensure that the calibration process is performed in accordance with the prescribed steps. The prescribed steps can be altered depending on the calibration task but also depending on intermediate measurement results. These aspects must be controlled with particular care. Alterations are

- positions of measurements used for the difference measurement,
- single or repeat measurements, conditions for repetitions,
- application or not of flattening correction,
- application or not of temperature correction.

Another sensitive aspect of the controlling software is the reliability under defined extreme conditions. This point is evidently a general one. Here we have the two points

- correct execution of measurements in accordance with the prescribed steps, also under extreme but permissible conditions, for example a temperature drift just below the allowed limit,
- rejection of execution or at least a warning in the case of inadmissible conditions or input data.

With respect to completeness of the software contents, risks of failures exist when the following functions are not supported sufficiently:

- input, use and archives of data sets of reference standards,
- input of parameters for the gauge block under test,
- control of the execution of measurements,
- print of test protocols, results and calibration certificates,
- archives of measurement results including a directory.

The measurement data processing part of the software performs the following functions:

- determination of length differences between the reference standard and the gauge block under test,

- determination of length differences between measurements in the center and in the corners,
- averaging of length differences determined,
- correction of flattening (addition of a corrective value),
- elimination of temperature influence (according to a given formula).

The only critical algorithmic point is the wipe-out of digits when the differences are calculated inappropriately.

Another crucial point of metrology software in general is the requirement of the repeatability of measurements. This means that, under the same conditions, the software must produce the same result. With gauge block calibrations, we have not observed any special problems concerning this point. But this important point should by all means be mentioned under the subheading "correctness with regard to the contents".

2.2 Correctness of the code

The source code of software is written by man. This is the reason why absolute correctness of the code cannot in general be guaranteed. Shortcomings or even systematic errors can not be excluded in the general case.

There are a lot of advanced tools to support software development or to test software, in particular with a view to detecting bugs. These tools - provided they are used properly - are a big help to reduce coding errors to a very low rate. However, there is still a residual error risk.

No particular risks are known with respect to the software for gauge block calibration as regards code correctness. What is generally true, also applies in this case.

2.3 Problems of data handling

Data handling has also proved to be a critical factor with respect to the accuracy of measurement results. Small changes of data caused by transformations can have a negative impact on the accuracy, for instance in connection with the wipe-out of digits. In addition to perturbations of data caused by numerical calculations, data can be perturbed by their transfer and necessary transformation on their path between different media, systems and men.

A gauge block calibration system has three interfaces and a general component which must be carefully considered. The first interface is that between the software system and the user. It realizes the input of data by the user, for example the input of the deviation of the reference gauge block from the nominal value. Sources of errors are

- mistakes or oversights of the user,
- transformation from the decimal to the dual representation, abbreviation or rounding errors.

The second interface is the take-over of data from the measurement hardware to the software. Sources of errors are

- the hardware interface itself,
- transformation of data between different forms of representation.

The third interface which causes possibly errors is that to external presentation of results on screens or in printed form. Here we have again the problem of

- transformation between different forms of representation.

A component of general importance for data correctness is the data storage system. Evidently, any mismatch in this part of the systems can lead to failures.

2.4 Proper use of software

Improper use or misuse of software is another crucial cause why systems can fail. Frequently, the users even do not become aware when it happens. There are two typical reasons for the improper use of software:

- incorrect or insufficient documentation of the software delivered, incorrect interpretation or incomplete study of documentation by the user,
- unstable behavior of the software and the users' attempt to change something on their own risk.

These points are generally true and they apply also to gauge block calibration.

In the gauge block calibration there are in particular a few points in the user interface identified which might cause at least misunderstandings:

- no strict separation of windows containing organizational data from windows related to the measurement process itself,
- no separation of windows handling parameters of the measurements from windows handling basic parameters of the measuring device,
- inconsistent and ambiguous usage of terms in different windows,
- sometimes missing references to appropriate standards during program execution (for example: concerning tolerances or algorithms).

2.5 Management problems

Software is a "living" product. It is changed from time to time for several reasons, for example to enhance the performance parameters, to correct defects, etc. Changes of software involve the risk of organizational mistakes. Because software is handled by different parties, problems such as incompatibility between pieces of software, invalid versions, incorrect documentation can arise. On the other hand, as software can be easily changed there is no natural barrier to prevent this.

As a consequence strong management rules should be established to prevent organizational mistakes. Important features are a complete identification of all those parts of the software that are subject to individual handling, the application of a unique version management, the assuring of correspondence between versions and their documentation.

If necessary, the supervision of an appropriate protection against manipulation of software or of data is another task incumbent upon the management.

3 Recommendations from the Software Technology Point of View

To minimize risks caused by software, the software engineering theory knows a variety of methods. In [4] we gave already a general overview about methods from software technology which might be useful in metrology. For further studies of such methods we recommend [1-3]. Here, in this paper, we want to refer only to a few working points which are in certain relations to the risks identified.

3.1. Program correctness and reliability

Correctness and reliability are fundamental requirements for software in general. To meet these requirements, two different kinds of software quality assurance measures are recommended: constructive measures for preventive avoidance of software defects and analytical measures aimed at the detection of defects and at software product evaluation. Software defects can be avoided to a large extent, when an application-dependent combination of the following elements is used in the software development process:

- a well-defined software development model (e.g. the life-cycle model of ISO/IEC 12207);
- a quality management system according to ISO 9001, ISO 9000-3;

- well-defined (standardized) development methods (structured analysis and design, programming guidelines, documentation rules, early test design, data base design);
- validated software tools if necessary.

Software product evaluation, and defect detection by software testing in particular, are basic techniques to increase the confidence in software products[1,2,3]. These techniques are supported by several international standards (e.g. ISO/IEC 14598, ISO/IEC 9126, ISO/IEC 12119). An appropriate combination of the following techniques should preferably be applied to verify calibration software:

- human testing (design and code inspections, code walkthroughs, reviews of the individual development phases),
- tool-supported static analysis of the non-executable software entity (code inspections, data-flow and control-flow analysis),
- tool-supported dynamic testing (black-box or functional testing based on program specification, white-box or structural testing based on the internal structure of the program, statistical testing based on a model of actual software usage).

3.2. Program identification and documentation

To support the handling of different programs and different versions of a program, each of the programs must be unambiguously identifiable. A minimal requirement is the indication of the program name, the version number and the manufacturer of the software (for example in the starting window). When a large number of programs or versions are handled, the application of configuration and change management tools is recommended.

The documentation is an essential part of each software product. The appropriate documentation must be available for each program. Quality requirements for software documentation have been defined in several international standards.

3.3. User interface

The part of the software that handles the user interface generally takes a significant percentage of the total system development time. The use of graphical user interfaces is often recommended. The windowing system should be derived from approved standards such as Windows or OSF. The direct manipulation interface should allow uniform menus, buttons and decision points (e.g. OK, Cancel, Help, Yes, No, ...) to be used. User inputs should be finished in a standardized form. A context-sensitive online-help facility should exist. The application of acoustic and/or optical signals is recommended for certain dangerous situations. At critical decision points in the user guidance process, the user should be requested to confirm the transactions.

3.4. Data handling and presentation of results

Recommendations for data handling comprise the domains of data input and transfer, data storage and data security. We emphasize that plausibility tests should be required for each data input. Above all the compliance of inputs with given parameter ranges and the unambiguity of identification numbers or names must be tested. Critical data must be protected against intentional and unintentional manipulations. Changes of critical data have to be confirmed before they are transferred to the data archive.

The data base of the calibration software must allow the correct and consistent archiving of data. Data must be organized in such a way that each of the performed processes is traceable. Specific data management utilities should be able to retrieve data selectively and to present results to the user. Utilities for data transfer from hard disk to floppy disk and for reconstruction of the data base from floppy disk are recommended.

The presentation of the results must be complete, unambiguous and consistent. The usage of terms must be consistent within the presentation as well as in relation to the program documentation as to the user interface. A subsequent presentation of the results must be possible. This means that a separation of the determination of the results from the presentation has to be realized.

4 Approaches for Calibration Services

After risks have been identified, the next important step is to propose practically applicable approaches with a view to a basically successful management of the calibration service. We have proposed two different approaches complementing one another. In this paper, we want to outline some facets of the first approach, which has largely been developed with respect to gauge block calibration. The second approach is only briefly sketched.

4.1 Checklist

Checklists are an approved means of the quality assurance. Basically, no additional technical equipment is required. However, they must cover sensitive aspects of the software.

With respect to gauge block calibration, it has turned out that working with a checklist representing a set of minimum mandatory requirements for the software is an appropriate means to enhance the reliability of software. A checklist can be used both when a calibration service is certified by external experts and when a self-check is made.

Prerequisite for the work with a checklist is:

- that an appropriate list of mandatory requirements is drawn up,
- that rules and instructions are formulated how to prove compliance with the requirements.

The elaboration of a set of mandatory requirements has been based on the analysis presented in section 2 and on general recommendations as presented in section 3. It has so far ended in a list of 37 requirements arranged in 7 different groups. The single groups concern

- documentation of software,
- identification of software,
- functions to fulfill,
- operating the software,
- data handling,
- manipulation of software,
- transparency and repeatability.

For illustration purposes, we list the single requirements of two groups. The first list consists of the 5 requirements of the group "operating the software":

(1) Important notions of the user surface as well as of the documentation are to be applied correctly, uniquely and consistently.

(2) The software must not break in any case.

(3) Each change of data has to be either stored permanently or rejected immediately. The software must not process data fed in incompletely.

(4) The change of stored, sensitive data has to be confirmed in addition.

(5) An error list of the software including recommendations for the continuation of the work must be available for the user.

The second list contains the 7 requirements of the group "Transparency and Repeatability":

(1) The software must store the measured values (raw data).
(2) The documentation must contain an exact description of the structure and access to raw data.
(3) Each set of raw data must be related to the position of the measurement.
(4) The software must deliver (on demand) a record containing mandatory information (for example: date of measurement, nominal measure, etc.).
(5) The print of a calibration certificate must be possible subsequently.
(6) Each modification of the software must be documented. Old versions must be kept for an appropriate period.
(7) The software must reproduce a known length difference between known reference standards within the uncertainty of measurement.

Concerning the examination of the requirements, instructions are necessary how to determine whether or not a concrete software complies with the requirements. For practical reasons, the instructions must be such that a non-specialist of software is able to understand and apply them.

The set of proving instructions established and arranged in instruction lines ensure in their entirety that all requirements are covered. There is no one-to-one correspondence between an instruction and a single requirement. Following the instruction lines, an examiner can check the compliance of the software with several requirements. This is illustrated by two examples whose significance concerning the contents could not be derived completely in this paper. In the examples, the underlined phrases represent the corresponding reaction on an existing requirement. The numbers count those requirements on which a proving instruction reacts. Not all of the requirements met by the instruction lines are listed above.

The proving instruction of the first example is as follows:
"Feed in a set of data for a reference standard.(1) Prove immediately after the input whether the data are usable for a difference measurement. (2) It must not be allowed. Instead, the software must demand to repeat or to confirm otherwise the input data. (3) If the software does so then repeat or confirm as demanded, store the data (4) and terminate. Start again the software and let show or print the data just fed in. Compare the showed data with the data fed in before (5). Pay attention to changes by cutting-off or by rounding.(6)"
This set of instructions proves altogether 6 requirements.
The second example consists of the following proving instructions:
"Start to feed in a second set of standard reference parameters. Try to give it an already existing identifier. (1) The software must reject to do this. Change the

identifier into a new one. Feed in a few parameters and break off before completing. Try to start a difference measurement using the incomplete data set of the reference standard. (2) The software must not allow to execute this measurement. Start again to feed in a data set with the last, valid identifier.(3) The previous values must not appear as available."

Here altogether 3 requirements are proved.

4.2 Standardized testing technology and test case design

The goal of software testing is to reveal faults in programs. Once an executable code has been written, it is possible to run test cases for this purpose. The design of relevant test cases (test data and expected results) is a fundamental task of software testing. However, there is no way to guarantee that all faults have been detected. Because of combinatorial explosion, too many test cases are to be consider in order that completeness can be achieved.

To guarantee the correctness and reliability of calibration software in a uniform way, the application of a standardized testing technology and the construction of standardized test cases have to be required. A standardized interpretation of test results would ensure the comparability of test results among all calibration services. When standardized test cases are run successfully, a uniform minimal level of software correctness is achieved.

There are two basic ways of constructing test cases for testing the correctness of calibration software. The first consists in designing test cases systematically according to the well-studied methods of software testing[1,2,3]. Another way to construct test cases is based on the experience gained by the calibration services from typical errors. Many errors of the last kind are concerned with the control of the measurement process. An example is the absence of a repeat measurement when specific requirements are violated.

The objective of systematic testing is to detect as many faults as possible. A reasonable strategy to pursue is to use at first black-box techniques (testing to specifications) and then to design additional test cases using white-box techniques (testing to source code).

A preparatory task to establish a standardized testing technology is the determination of relevant testing areas in the calibration process. Examples of such areas are the measurement evaluation, the data management process, protection against manipulation, and controlling of different kinds of calibration (for example with regard to different standards). In the following, some aspects are considered within the scope of test cases for processing the measured length differences of gauge blocks.

Using black-box techniques, test cases have to be constructed considering the different parameters of the computation algorithm. These parameters include

- kind of measurement: single measurement or multiple sensor touching with averaging,
- measurement position: five positions at gauge block (computation of the total deviation range) or only center position (computation of the deviation range for the center position),
- computation with flattening correction or not,
- computation with temperature correction or not.

Sixteen classes of test cases have to be derived to cover all possible parameter combinations. Because of the independence of the parameters, the following subset of classes of test cases seems to be sufficient to test the correctness of the computation algorithm:

- Testing of averaging: multiple measurement at central position, standard probe and gauge block under test consist of the same material, no flattening correction, no temperature drift.
- Testing the computation of the deviation range: single measurement at five positions, standard probe and gauge block under test consist of the same material, no flattening correction, no temperature drift.
- Testing the flattening correction: single measurement at central position, standard probe and gauge block under test consist of different materials, no temperature drift.
- Testing the temperature correction: single measurement at central position, standard probe and gauge block under test consist of the same material, no flattening correction, significant temperature drift during measurement.

These test cases are relevant if the original raw data (measured length differences of gauge blocks) are plausible and thus transferred to the computation algorithm. Other test cases have to be developed for the following situations:

- raw data are rejected because of unsuccessful plausibility tests or exceeding of the measuring range of the device;
- high scattering of raw data, stopping the evaluation or displaying a warning;
- raw data are evaluated but inadmissible changes in temperature occur; stopping the evaluation or displaying a warning.

There must be agreement between the results of the test cases run and the results of the manual evaluation of the original raw data.

5 Conclusions

The investigations have turned out that in the case of gauge block calibration a systematic software risks analysis and based on it an elaboration of quality measures are possible.

The demonstrated approach is repeatable for many other cases. This approach is even useful if the investigations cannot be completed in a first attempt. The results may be used already in any intermediate stage, and they can be refined step by step.

Based on the investigation of one case, it is not possible to assess to what extend the results are transferable to other cases. However, the lessons that can be learned from many case studies may lead to more general methods.

Acknowledgment

This study was stimulated by the German Calibration Service, Technical Committee for Length Measurement. The authors are much obliged for support and advice of the PTB Laboratory for Interferential Length Measurement headed by Dr. Gerhard Bönsch.

References

1. B. Beizer, Software Testing Techniques, (Van Nostrand, New York, 1990).
2. B. Beizer, Black-Box Testing, (John Wiley, New York, 1995).
3. G.J. Meyers, The Art of Software Testing, (John Wiley, New York, 1979).
4. D. Richter, Software Quality Assurance in Metrology, In Advanced Mathematical Tools in Metrology, ed. P. Ciarlini et.al., (World Scientific, Singapore, 1994).
5. Guidelines for the Development of Software to be Used in Test and Measuring Laboratories, Nordtest Technical Report 145, (ESBO, Finland, 1991).

Advanced Mathematical Tools in Metrology III
Edited by P. Ciarlini, M.G. Cox, F. Pavese & D. Richter

ROBUST ALTERNATIVES TO LEAST SQUARES

WERNER STAHEL

Swiss Federal Institute of Technology (ETH) Zürich, CH-8092 Zürich, Switz.

stahel@stat.math.ethz.ch

Least Squares estimation in regression models is well known to be optimal if the random errors are normally distributed. In the presence of longer-tailed errors or outliers, there are better alternatives, called robust estimators.

Two robustness measures are based on the idea of gross errors – arbitrary observations added to a given sample. The gross error sensitivity measures the maximal effect of a single observation, and the breakdown point characterizes the fraction of gross errors that leads to arbitrarily large effects on the estimate.

M-estimators are a flexible class of estimators suitable for obtaining robust procedures with desired properties. In the regression context, they show only limited and often unsatisfactory robustness. On the other hand, desired robustness properties can only be achieved at the cost of heavy computation.

These concepts are explained on the basis of two examples from environmental sciences and molecular spectroscopy. In the latter, robust methods were essential for finding a useful solution.

1 Introduction

Regression and Least Squares. Fitting a model to data is the common theme of a large part of statistical tasks in science and technology, and certainly a basic topic in metrology. The methods most often used are based on the *Least Squares* (LS) principle. They play this fundamental role because the principle is simple, they are relatively easy and cheap to compute, and they perform optimally under the assumption of normal distributions of the random quantities. However, if the normality assumption is not valid, there exist alternative methods which beat these techniques – by far in some cases.

Robust Statistics. The field of Robust Statistics studies, in general, the effects of deviations from model assumptions and provides methods which still work well if the assumptions are met only approximately.

Goal. This paper deals mainly with deviations from the assumed normal distribution of the random errors in multiple linear regression and introduces some basic concepts in robustness. A related, nice introduction has already been given by Vecchia and Splett [1] in the same context. While there is some necessary overlap, a more extended overview is given here, and two further applications illustrate the success of robust procedures. An outlook on other topics in robustness concludes the paper.

2 Multiple Linear Regression

Notation. The general model for (multiple) linear regression will be denoted as

$$Y_i = \sum_j x_i^{(j)} \beta_j + E_i = \underline{x}_i^T \underline{\beta} + E_i \ . \tag{1}$$

Here, Y_i is the value of the *target variable* for the ith observation, and $\underline{x}_i = [x_i^{(1)}, ..., x_i^{(m)}]^T$ is the (column) vector of respective values for the *explanatory variables* (or design variables or carriers). Usually, one explanatory variable $x^{(j)}$ will be identically 1, such that the corresponding coefficient β_j acts as the "intercept". E_i is the random error. The usual model assumes that the E_i are independently, identically and normally distributed, $E_i \sim G_r = N(0, \sigma^2)$. The parameters are thus the coefficients β_j, collected into the vector $\underline{\beta}$, and the scale σ of the errors.

Tunnel measurements of air pollution. Road traffic is an important source of air pollution. In order to quantify its contributions, it is necessary to obtain "emission factors", that is, to estimate the average quantity of a given pollutant like NO_x which a vehicle emits per kilometer. In a one-way tunnel, the total pollution caused by traffic in 15 minutes, say, can be measured together with the number of vehicles, and an average emission factor Y_i can be calculated. If the vehicles are classified into different categories, an emission factor for each category can be obtained: Figure 1 shows average emission factor values together with the proportion x_i of heavy duty vehicles for a study in a tunnel near Zurich (see [2] for details). Since heavy duty vehicles emit more NO_x than passenger cars, Y_i increases with x_i. In fact, since the total pollution is the sum of the contributions from light and heavy duty vehicles, it is straightforward to postulate the simple regression model

$$Y_i = \alpha(1 - x_i) + \gamma x_i + E_i = \alpha + \beta x_i + E_i \ ,$$

where α is the emission factor for light duty vehicles and $\gamma = \alpha + \beta$, for heavy ones.

A Least Squares fit is shown in Figure 1. Clearly, the distribution of the errors is skewed and depends on the explanatory variable x. One wonders if a different straight line might give more adequate results for the two important parameters α and γ.

An example from Molecular Spectroscopy. Insight into the physical structure of atoms and molecules can be obtained by assessing the pattern of energy levels associated with their quantum mechanical states. Experiments associated with this topic form the scope of *atomic and molecular spectroscopy*. The energy levels are obtained by measuring the energies connected to the

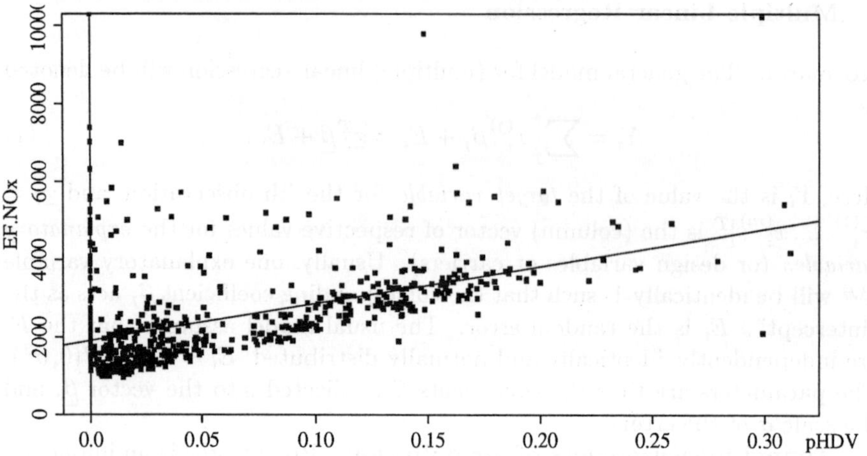

Figure 1: LS regression of the average emission factor against the proportion of Heavy Duty Vehicles.

transitions of a molecule between different quantum mechanical states. The corresponding model is extremely simple and reads

$$Y_{\ell u} = -\theta_\ell + \theta'_u + E_{\ell u}\,, \tag{2}$$

where $Y_{\ell u}$ is the energy released when the molecule falls from the *upper state* u to the *lower state* ℓ, and $E_{\ell u}$ is the random measurement error. The quantities θ_ℓ and θ'_u are the unknown energy levels of the two states. Since only differences are measured, the levels can be determined only up to an additive constant. The energy level of the lowest θ_ℓ is therefore set to zero.

The data set which will be analyzed here consists of 659 measured transitions involving 249 quantum mechanical states of the tritium molecule T_2, which can be separated into 54 lower and 195 upper states. (See [3] for a more complete description.)

The model can be rewritten in the form of a multiple regression model, $Y_i = \sum_j x_i^{(j)}\theta_j + E_i$ using

$$x_i^{(j)} = \begin{cases} 1 & \text{if } j \text{ is the upper state for the } i\text{th observation,} \\ -1 & \text{if } j \text{ is the lower state for the } i\text{th observation,} \\ 0 & \text{otherwise.} \end{cases}$$

Each $\underline{x}_i$ thus contains a 1, a -1, and 246 zeros (except for those transitions involving the lowest state). Such a situation is called a *sparse design*.

The measurement errors E_i are extremely small. Since molecules cannot be forced to switch between selected states, highly sophisticated methods are used to determine a pair $[u_i, \ell_i]$ for each measurement Y_i, and this process may partly fail. Mistakes lead to outliers of arbitrary size with respect to the given model. We shall show that the LS method leads to useless results, while a robust alternative yields adequate estimates of the parameters and allows for a generally reliable identification of the falsely identified pairs of states.

Are random errors normal? Even though the mechanism leading to gross errors is very special in this example, practitioners know that *outliers* do occur. Usual practice suggests to search for outliers informally or by formal tests, find the reason for the blunders, correct or remove the observations in question, and use LS on the rest.

Metrology is often concerned with high-quality data, and outliers rarely cause a real problem. On the other hand, random measurement errors are usually not normally distributed, but they show *longer tails.* This statement is difficult to verify, since some of the common goodness-of-fit tests have resticted power and samples are too small. In large samples of high-quality data, long tails become evident (see [4], p. 22).

Least Squares. It is generally known that

- the Gauss-Markov theorem proves that the LS estimators have minimal variance also for non-normal errors (as long as the variance of the E_i exists) within the class of *linear* estimators (linear in the observations Y_i). – Robust alternatives are non-linear (as is outlier rejection followed by LS). They typically show much *smaller standard errors* (are statistically more efficient) for long-tailed error distributions. For strictly normal random errors, they are necessarily less precise than LS, but this loss can be kept small. In terms of an insurance, a small premium (efficiency loss for normal errors) is sufficient for a large protection (against efficiency loss for longer-tailed distributions).
- t- and F-tests based on LS are adequate as long as the errors are approximately normal or there are many more observations than parameters in a reasonably balanced design. – While this is true for their validity (error of the first kind), alternatives usually have better power (smaller probability of errors of the second kind) and thus lead to shorter confidence intervals for long-tailed error distributions.

There is a large variety of alternative, robust methods (see [4], Sec. 2.3b, and [5]). Here, we want to present only two basic classes.

3 M- and GM-estimators

Definition. A first class of alternative estimators is obtained by slightly changing the criterion minimized by Least Squares: An *M-estimator* for model (1) is the value $\underline{\beta}$ which minimizes

$$\sum_i \rho(r_i/\widehat{\sigma}) \,, \tag{3}$$

where $\widehat{\sigma}$ is a suitable estimator of the spread of the random errors, and ρ is a suitable scalar function. In practice, $\widehat{\sigma}$ usually needs to be obtained from the residuals, which in turn depend on β. If $\rho(r) = r^2$, the LS estimator results.

What functions ρ are suitable? If we knew that the errors followed a known long-tailed distribution with density f, we would be led to using the corresponding maximum likelihood estimator, since it is asymptotically most efficient. In fact, this is an M-estimator with ρ-function $\rho(r) = -\log f(r)$. If f is the normal density, the LS estimate is obtained. Figure 2 shows a few useful ρ-functions and their derivatives $\psi = \rho'$.

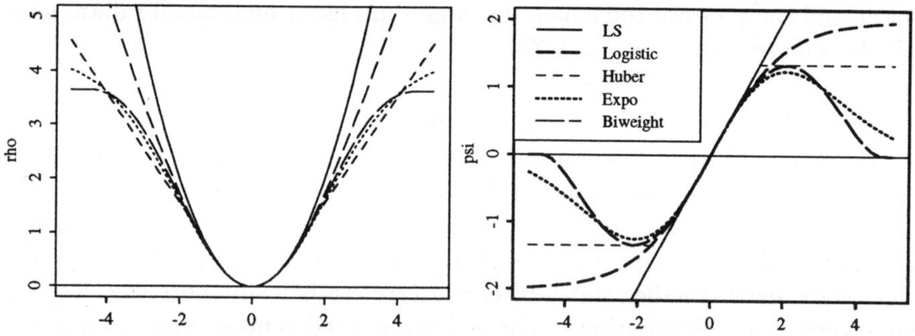

Figure 2: Five useful ρ-functions (left panel) and their derivative ψ (right panel)

In order to solve the problem of minimizing (3), the derivatives are set equal to zero, which leads to the so-called normal equations

$$\sum_i \psi([Y_i - \underline{x}_i^T \underline{\beta}]/\widehat{\sigma})\, x_i^{(j)} = 0 \tag{4}$$

for all j, with $\psi = \rho'$. This helps to devise both theoretical results and algorithms. As we shall see shortly, the function ψ plays a more prominent role than ρ.

A more general class allows the ψ-function to depend on $\underline{x}$: A *GM-estimator* is defined as the solution of

$$\sum_i \eta(\underline{x}_i, [Y_i - \underline{x}_i^T \underline{\beta}]/\widehat{\sigma})\, x_i^{(j)} = 0 \,, \qquad (5)$$

where η is usually of the form $\eta(\underline{x}, r) = \eta(\|\boldsymbol{A}\underline{x}\|, r)$ for a suitable matrix $\boldsymbol{A}$ which depends on the design $\underline{x}_1, ..., \underline{x}_n$.

Their Use. In applications, the differences between LS and robust (point) estimates are often rather small. This relates to the general fact that even considerable efficiency losses are difficult to grasp on the basis of single cases.

In practice therefore, robust estimates are often used only as a *diagnostic tool*: Both the LS and a robust estimate is obtained. If they show reasonable agreement, the LS results are reported, because they are not questioned by referees and readers. If larger discrepancies appear, they can often be traced down to a few extreme observations. These outliers may then be rejected, and the LS results for the remaining observations are reported. Clearly, tests and confidence intervals based on such a streamlined sample are, in principle, invalid. The strategy is a version of formal or informal *outlier rejection* procedures. Rejection rules are usually based on a LS fit for the whole data set. It can be shown that rejection of outliers based on a robust fit instead is more reliable.

Results. For the *tunnel example*, the results of a robust fit are compared to the LS results in Figure 3. In this study, a reason could be found for the many observations with unexpectedly high Y values: When there was little traffic, the air flow was not strong and steady enough for the calculation of meaningful average emission factors Y_i for 15 minutes. The LS estimate (and the robust one, not shown) for the observations with sufficient air velocity agrees reasonably with the robust estimate based on all the data, but is rather different from the corresponding LS estimate.

Of course, in such a case, where reasons for outliers are available and the model can easily be visualized, any reasonable analysis will find a reasonable result. Robust estimators are needed more urgently in more complex situations.

In the Molecular Spectroscopy example, we applied both LS and the M-estimator based on the logistic ψ-function, using the MAD of the residuals as a scale estimate. The LS residuals show a clearly long-tailed distribution with some 4% extreme outliers (Figure 4). Their median absolute value (MAD) is 80, but the measurement error is known to have a standard deviation of about 0.02 units (wave numbers, cm^{-1})! As a consequence of the sparse design, there are 49 exactly zero residuals. All the other residuals are larger than 0.1!

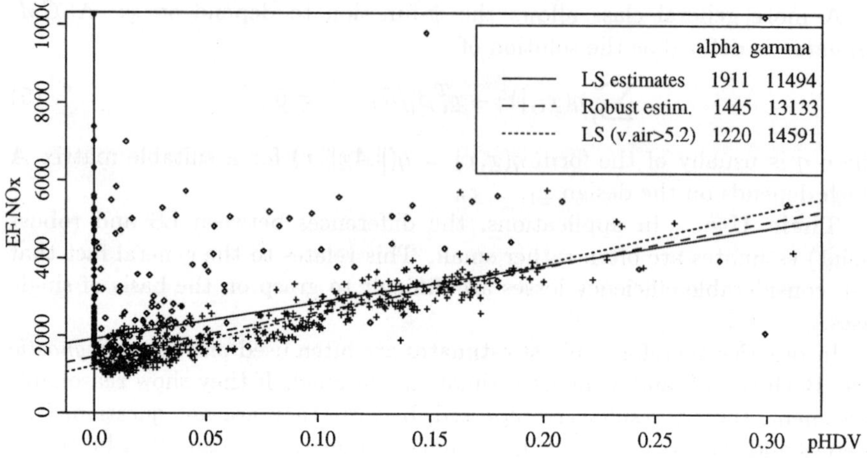

Figure 3: Three regression lines for the tunnel data: LS, robust, and LS applied to the data with enough air flow. The latter are marked with crosses, the rest, with diamonds

For the robust fit, only about 1/3 of the residuals are larger than this value, the MAD being 0.021 (corresponding to a standard deviation of 0.03). They can be regarded as cases of wrong identification of lower or upper quantum mechanical states. Furthermore, the estimated energy levels should show a certain structure with respect to their quantum numbers according to physical theory. The pattern of the robust estimates fits well in contrast to the LS estimates. See [3] for details.

4 Robustness Measures

In order to compare the robustness of different estimators, criteria which describe their reaction to the violations of model assumptions are needed. A procedure is *(locally) robust if small deviations from the assumptions lead to small changes in the results.* It is *globally robust* or *resistant* if the results remain reasonable even for rather large deviations from the assumptions.

For studying the local behavior, we calculate the change in the estimate caused by adding an observation at $[\underline{x}, y]$ to the sample $[\underline{x}_1, y_1], \ldots, [\underline{x}_n, y_n]$ and obtain the Sensitivity Curve (or function)

$$\begin{aligned}&\mathrm{SC}(\underline{x}, y; [\underline{x}_1, y_1], \ldots, [\underline{x}_n, y_n]) \\ &\qquad = n\left(\widehat{\underline{\beta}}([\underline{x}_1, y_1], \ldots, [\underline{x}_n, y_n], [\underline{x}, y]) - \widehat{\underline{\beta}}([\underline{x}_1, y_1], \ldots, [\underline{x}_n, y_n])\right).\end{aligned}$$

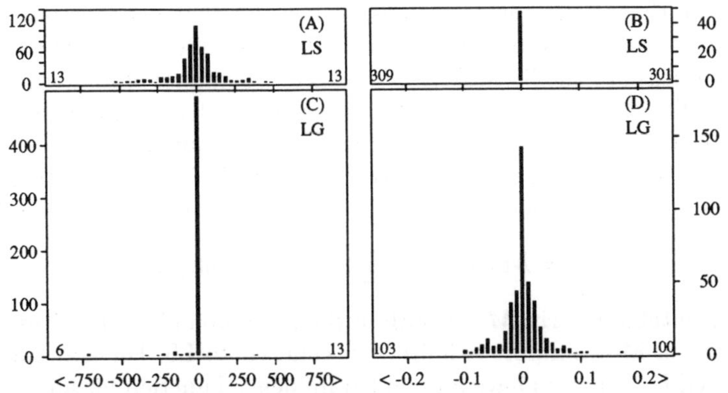

Figure 4: Distribution of residuals from LS (A,B) and a robust fit (C,D). The histograms show the residual in 2 different ranges (A,C and B,D). The numbers of residuals falling outside the ranges are noted next to the histograms.

Figure 5 shows the effect of adding an observation with $x = 0.34$ as a function of a variable y for the slope's M-estimators given by the ρ- and ψ-functions in Figure 2. They have a similar shape as the ψ-functions.

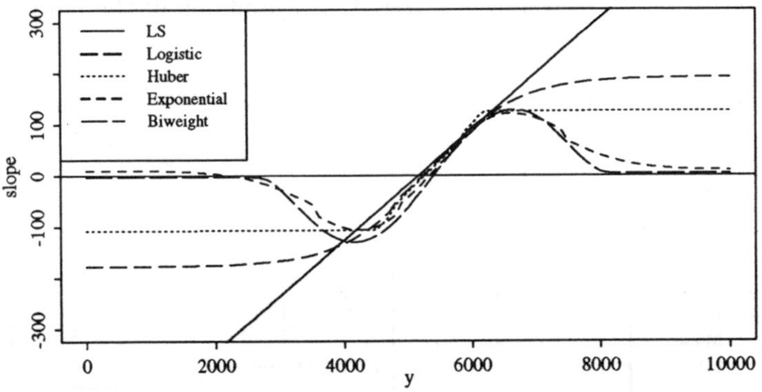

Figure 5: Sensitivity Curves: Estimated slope as a function of the y coordinate of an additional observation at $x = 0.34$ in the tunnel example

Since, in general, the SC is vector valued like $\underline{\widehat{\beta}}$, a suitable norm $||.||$ is used to quantify the effect, and its maximum

$$\max ||\mathrm{SC}(.,.;[\underline{x}_1, y_1], ..., [\underline{x}_n, y_n])||$$

is a useful measure of robustness.

For theoretical considerations, a measure that depends on the assumed model rather than a sample is more handy. The Influence Function corresponds in this sense to the Sensitivity Curve. The maximum of its norm is the *Gross Error Sensitivity*. It measures the worst effect that a "small contamination" can have on the estimate, and is therefore a measure of local robustness.

For M- (or GM-) estimators, the Influence Function is exactly proportional to the ψ- (or η-) function and to a suitable linear transformation of $\underline{x}$,

$$\text{IF} = \boldsymbol{M}\underline{x}\, \psi(y - \underline{x}^T\underline{\beta}) \quad (= \boldsymbol{M}\underline{x}\, \eta(\underline{x}, y - \underline{x}^T\underline{\beta})\ )$$

for a suitable matrix $\boldsymbol{M}$. If both of them are bounded, the Gross Error Sensitivity of M-estimators is finite – in contrast to LS. For appropriate choices of η, GM-estimators have a bounded Influence Function for any design. They are called *Bounded-Influence Estimators*, see [4], Chap. 6.

Global Robustness. In order to study a "global" aspect of robustness, we now play the game of altering a fixed number m of observations rather than adding a single one: Let $[\underline{x}_1^*, y_1^*], ..., [\underline{x}_n^*, y_n^*]$ be a sample for which $n - m$ observations are identical to observations in the given sample $[\underline{x}_1, y_1], ..., [\underline{x}_n, y_n]$, while the other m observations are arbitrary. In the tunnel example, we may change the last $m = 1, 2, 3, ...$ observations to be equal to the outlier $[x = 0.3, y = 10,215]$ in the upper right hand corner of Figure 1. Figure 6 shows three estimates of α and $\gamma = \alpha + \beta$, as a function of the number m of faked data.

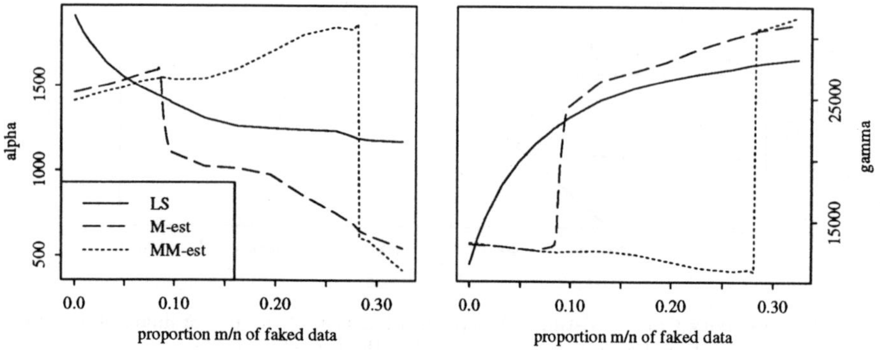

Figure 6: Estimates $\widehat{\alpha}$ and $\widehat{\gamma} = \widehat{\alpha} + \widehat{\beta}$ as a function of the number m of observations [pHDV=0.3,y=10,215] added

Now examine the worst effect of such alterations on the estimator! If any component of $\widehat{\underline{\beta}}$ can be made arbitrarily large by "wicked" choice of the faked observations, we say that *the estimator breaks down.* The largest proportion $(m-1)/n$ for which this does not occur is the *finite-sample breakdown point.* Usually, it is essentially independent of the sample size *and* of the "good" observations $[\underline{x}_1, y_1], ..., [\underline{x}_n, y_n]$ (unless they show a special configuration). As with local robustness, there is an "asymptotic" version of the concept.

In our *Molecular Spectroscopy example*, it seems clear at first glance that the robust procedure has not broken down. Formally, however, this is not true, since some components of the parameter vector appear wrong in the light of their clear mismatch with the physical theory mentioned above. The reason is easily understood: Some of the energy levels θ_u of upper quantum mechanical states are connected only to very few observed transition energies $Y_{\ell u}$. If half of these are modified appropriately, the estimate is misled to assume an arbitrary wrong value. This phenomenon, which is a consequence of the sparse design, is called *partial breakdown*, see [6] and [7].

The Breakdown of GM-Estimators was found to be bounded by $1/p$ (in case of random explanatory variables $\underline{x}_i$), where p is the number of parameters β_j. That is, for 7 explanatory variables, any M-estimator breaks down if we are allowed to manipulate 1/7=14% freely. Now, this seems to be a rather high fraction of gross errors that can be tolerated. Note, however, that an observation $[\underline{x}_i, y_i]$ becomes useless in principle if only one of its components is grossly wrong. It only takes a fraction of 2% of false values to make the probability of at least 1 false in 8 values equal to 1/7 – and 2% is a rather low fraction of gross errors in practice!

The bound $1/p$ applies to GM-estimators defined as the solution of the normal equations. If there is more than one solution, they usually correspond to local minima of the corresponding criterion (3). Then, the result says that at least one solution breaks down for contamination fractions $\geq 1/p$. There may still be an adequate solution, and the next section describes how to find it.

5 S- and MM-estimators

S-estimators. A second class of estimators, besides the M-estimators, has been found to show good breakdown properties. The idea of Least Squares is to minimize the variance of the residuals. Since the variance is a non-robust measure of the spread, it is natural to replace it by a robust measure $\widehat{\sigma}(r_1, ..., r_n)$. Thus, an *S-estimator* of regression is given by

$$\widehat{\underline{\beta}} = \arg\min_{\underline{\beta}} \widehat{\sigma}(y_1 - \underline{x}_1^T \underline{\beta}, ..., y_n - \underline{x}_n^T \underline{\beta})$$

The most well-known instance is the Least Median of Squares method of Hampel and Rousseeuw ([8]), which minimizes median(r_i^2).

The function to be minimized is not convex for robust members of this class. Therefore, calculation of the minimizing $\underline{\beta}$ is extremely difficult for more than 2- or 3-dimensional arguments $\underline{\beta}$. A stochastic algorithm is used which finds the solution with some desired probability ([9]). If this number should be high for a high-dimensional problem, the computer time needed is exorbitant. In practice, a compromise is needed, and the high protection from breakdown often remains a hope.

A second handicap of these estimators is their statistical inefficiency: Estimators with a high breakdown point are much more variable than good M-estimators.

MM-estimators. Both high breakdown point and high efficiency can be obtained by combining S- and M-estimation, leading to MM-estimators. These proceed in two steps:
Step 1: Calculate an S-estimate $\widehat{\underline{\beta}}_0$ (by the probabilistic algorithm mentioned above, followed by a local minimization).
Step 2: Find the local minimum of the criterion (3) of an M-estimator which is "nearest" to $\widehat{\underline{\beta}}_0$ by performing an iterative algorithm for solving the normal -2pt] equations (4), starting from $\widehat{\underline{\beta}}_0$.
A detailed proposal is given by Yohai et al. [10] based on [11].

6 Tests and Confidence Regions

Asymptotic Inference. Point estimates are of limited practical use unless they are accompanied by testing procedures and confidence intervals. To this end, the distributions of the estimates need to be known. For M-estimators, asymptotic results are well-known and quite simple to understand: They are asymptotically normal with covariance matrix $\sigma^2\tau\boldsymbol{C}$, where $\boldsymbol{C} = (\sum_i \underline{x}_i\underline{x}_i^T)/n$ and σ is the spread (standard deviation) of the error distribution G_r as for LS, but a "correction factor" $\tau = \int \psi(r/\sigma)^2 dG_r(r)/(\int \psi'(r/\sigma)dG_r(r))^2$ is needed.

The covariance matrix of the estimates can thus be estimated by $\widehat{\boldsymbol{V}} = (\widehat{\sigma}^2/n)\widehat{\tau}\widehat{\boldsymbol{C}}$, where

$$\widehat{\tau} = \frac{\text{ave}_i[\psi(\widetilde{r}_i)^2]}{[\text{ave}_i\psi'(\widetilde{r}_i)]^2}, \qquad \widehat{\boldsymbol{C}} = \frac{\text{ave}_i\underline{x}_i\underline{x}_i^T w_i}{\text{ave}_i w_i}, \tag{6}$$

$\widetilde{r}_i$ are suitably standardized residuals, like $\widetilde{r}_i = (y_i - \underline{x}_i\widehat{\underline{\beta}})/\widehat{\sigma}$, and $w_i = 1$ or, alternatively, $w_i = \psi(\widetilde{r}_i)/\widetilde{r}_i$ as proposed in [10]. For all of these factors of $\widehat{\boldsymbol{V}}$ alternative, asymptotically equivalent estimators can be proposed.

For finite samples, the asymptotic approximation may be insufficient, and the alternatives may behave rather differently. In his fundamental paper, Huber [12] introduced a further "finite sample correction" factor. Very little is known about the merits of such alternatives, including bootstrap procedures.

Is Inference for LS exact? The need for asymptotic approximations appears to be a serious drawback of robust estimators. The distribution of LS is well-known to be exact for finite samples – *if the distribution of the errors is normal* and the variances equal (or correctly specified). If this myth is left behind, there is no reason to believe that we are better off. Rather, there are arguments which suggest that *the normal distribution approximates better for robust estimates than for LS!* (Assume a stable law with shape parameter $\alpha < 2$ for E_i. Then, the LS estimate has a stable distribution, too, and its convergence rate is slower than $\sqrt{n}$, whereas robust estimates converge at the usual speed to a normal distribution.)

7 Outlook on other Robustness Problems

In this paper, we have explained and illustrated some concepts and methods of robust statistics for the model of multiple linear regression. Similar ideas can be and are applied in other areas:

Heteroskedasticity. The assumption of equal variance for all the random errors (or a known pattern of variances) can also be violated, and procedures which are robust in this sense have been studied. To some extent, the robust estimators discussed above solve this problem, too, since an E_i with a large $\text{var}(E_i)$ will, with high probability, appear as an outlier, and its influence will be reduced.

Non-Linearity. If the assumed linear model $\underline{x}^T\underline{\beta}$ for the conditional expectation $E(Y|\underline{x})$ is wrong, the problem is of a quite different nature. It is then unclear what the "ideal" result would be, that is, to which parameter values an estimator should converge. The use of robust estimators can lead to worse results in such a context: Consider the situation of a single explanatory variable for which the true model should include a quadratic term, but the postulated model is linear. As long as we are interested in an approximate description or in interpolation, all estimators may have their range where they lead to most adequate results. As soon as extrapolations are needed, robust estimators may be worse than LS: They allow for some points with a bad fit, and these will often be the observations with extreme $\underline{x}$'s in such a situation. The fit beyond these extreme points may even be worse.

Dependence Structure. For tests and confidence intervals to possess the assumed error or coverage probabilities, independence of the random errors is essential (or, more generally, the assumed correlation structure must be correct). Deviations from this assumption are difficult to study in any generality, since there are too many types. Some work has been done for the case of a sequential structure in time. For the case of autocorrelations of the random errors, there is a classical remedy: First, obtain a preliminary fit ignoring the dependence. Then, derive an AR or ARMA time series model for the residuals. Finally, use the general least squares theory to obtain improved estimates and/or correct tests and confidence intervals.

The most dangerous deviation from independence, however, is a *long-range dependence*, which cannot be adequately described by the ARMA models. Such correlations can have disastrous effects for longer series. What's more, they have been shown to occur frequently in real data sets ([4], Chap. 8.1) – even in series of chemical and physical measurements where there is no plausible reason for dependence. Again, the the correlation structure should be modeled and the corresponding inference procedures used. Künsch et al. [13] study these questions and find, reassuringly, that if contrasts between groups or slope parameters in regression models are estimated and the design variables are in some sense orthogonal to time, the levels of tests are correct.

Other models. M-estimation has been extended to *non-linear* and to *generalized linear models.* Random effects and mixed models have only recently been treated. The estimation of a "scatter matrix" (covariance matrix) forms the basis for multivariate statistics and has received considerable attention in the robustness literature. The approach in [4] can be applied to any parametric model for independent observations. For time series, the concepts and methods need some extension, see [4], Chap. 8.3.

A taxonomy of these and other research directions is given in [14].

8 Software

Robust methods are not yet widely available in software packages apart from robust measures of location and scale for a univariate sample.

The system which provides the largest coverage is the package S and its extension S-Plus ([15], [16], [17], [18]). It implements a powerful language with many useful statistical, graphical and mathematical functions and allows for manipulating data and combining it with results of analyses in very flexible ways. Routines written in C or Fortran can be incorporated. It has become a language of communication among statisticians, and a large collection of contributed procedures, called statlib, is available on the internet.

The standard S/S-Plus package presently includes several robust estimation procedures for M-estimation (rreg), L_1-regression (l1fit), the two S-estimators Least Median of Squares (lmsfit) and Least Trimmed Squares (ltsfit), and finally a kind of robust M-estimation method (robust) for ordinary and generalized linear models (which, however, is not really robust since the scale is estimated non-robustly). Tests are provided only for the last procedure. They are poorly documented.

In the "statlib" collection, there is a function for robust M-estimators (rlm) which also provides tests in the library MASS by Venables and Ripley [17] and a function (mmreg) for the MM-estimator proposed in [10] and mentioned above.

A large collection of routines implementing many robust methods has been developed by Marazzi [19] and is called ROBETH. There is an interface which allows to use all these routines under the S system. Its use requires some expertise, however.

Standard packages do not provide robust regression methods. SAS documents how the procedure NLIN can be used to provide M-estimates, but the standard errors and tests given in the output are inadequate.

9 Concluding Remarks

It has been demonstrated that Least Squares methods are rather unsafe, since they give non-optimal and sometimes misleading results in situations which often do occur in practice. If diagnostics are used to verify the assumptions, it is often (but not always) possible to detect serious violations, but remedies in the form of recipes have unknown statistical properties. Therefore, it is clearly safer to use robust estimators from the start.

Such a statement is often misunderstood. It may seem that the use of robust methods avoids the need for deriving an adequate model. Of course, this would be a fundamentally wrong attitude: There is no hope to obtain good estimates and correct inference for inadequate models! In contrast:

- Suitable robust fitting methods show deviations from the assumptions more clearly than LS, and therefore chances that a suitable model is found increase. (I must admit that this statement is subject to an ongoing discussion.)
- No model ever is exactly "correct" – except for simulation studies and similar artificial situations. The model should be a good approximation, and if this is assumed, robust methods are better than "optimal" classical methods.

Why, then, are robust methods not more popular?

- Deviations, even if only the distribution is considered, come in several flavors (few or many outliers, in the E_i only or in $\underline{x}_i$, too), which lead to different corresponding procedures.
- There have been many approaches to these problems. Finding new ramifications is more rewarding in an academic environment than "consolidating" a specific approach.
- Most research has been devoted to estimation. Users want tests and confidence intervals which go along with them. Asymptotic results have been available for 25 years now. But, in contrast to other fields like generalized linear models, robustness experts have hesitated to advocate corresponding inference procedures.
- Software package writers need one or two "general purpose" estimation and inference procedures, which should be generally accepted by the "robustness community" and can be presented in a form which is understandable for common users. They need a partner with enough knowledge and energy who can turn a statistical method into a "consumer product". S/S-Plus has made a start here.
- Convincing examples should make the need for such methods clear to the potential users. Rousseeuw [20] has collected a number of engineering applications. We hope that this paper also contributes to this end.

References

1. D. F. Vecchia and J. D. Splett. Outlier-resistant methods for estimation and model fitting. In P. Ciarlini, M. G. Cox, R. Monaco, and F. Pavese, editors, *Advanced Mathematical Tools in Metrology*, pages 143–154. World Scientific, Singapore, 1994.
2. J. Staehelin, K. Schläpfer, T. Bürgin, U. Steinemann, S. Schneider, D. Brunner, M. Bäumle, M. Meier, C. Zahner, S. Keiser, C. Keller, and W. A. Stahel. Emission factors from road traffic from a tunnel study (Gubrist tunnel, Switzerland). Part i: Concept and first results. *Science Total Environment*, 169:141–147, 1995.
3. Werner A. Stahel, Andreas F. Ruckstuhl, Peter Senn, and Kurt Dressler. Robust estimation in the analysis of complex molecular spectra. *J. Am. Statist. Assoc.*, 89(427):788–795, 1994.
4. F. R. Hampel, E. M. Ronchetti, P. J. Rousseeuw, and W. A. Stahel. *Robust Statistics: The Approach Based on Influence Functions.* Wiley, N. Y., 1986.

5. Werner A. Stahel and Sanford Weisberg. *Directions in Robust Statistics and Diagnostics*, volume 1, 2. Springer, N. Y., 1991.
6. F. R. Hampel. Some problems in statistics. In Yu. Prohorov and V. V. Sazonov, editors, *Proc. 1st World Congress of the Bernoulli Soc.*, volume 2, pages 253–256. VNU Science Press, Utrecht, 1987.
7. Andreas F. Ruckstuhl. Partial breakdown in two-factor models. *J. Statist. Planning Inference*, 1996.
8. P. J. Rousseeuw. Least median of squares regression. *J. Am. Statist. Assoc.*, 79:871–880, 1984.
9. P. J. Rousseeuw and A. M. Leroy. *Robust Regression & Outlier Detection.* Wiley, N. Y., 1987.
10. V. Yohai, R. Zamar, and W. A. Stahel. A procedure for robust estimation and inference in linear regression. In W. A. Stahel and S. Weisberg, editors, *Directions in Robust Statistics and Diagnostics*, volume 2, N. Y., 1991. Springer.
11. V. J. Yohai. High breakdown-point and high efficiency robust estimates for regression. *Ann. Statist.*, 15(2):642–656, 1987.
12. P. J. Huber. Robust statistics: A review. *Ann. Math. Statist.*, 43:1041–1067, 1972.
13. H. R. Künsch, J. Beran, and F. R. Hampel. Contrasts under long-range correlations. *Ann. Statist.*, 21(2):943–964, 1993.
14. W. A. Stahel. Research directions in robust statistics. In W. A. Stahel and S. Weisberg, editors, *Directions in Robust Statistics and Diagnostics*, volume 2, NY, 1991. Springer.
15. Richard A. Becker, John M. Chambers, and Allan R. Wilks. *The new S Language; A Programming Environment for Data Analysis and Graphics.* Wadsworth & Brooks/Cole, Pacific Grove, 1988.
16. John M. Chambers and Trevor J. Hastie. *Statistical Models in S.* Wadsworth & Brooks/Cole, Pacific Grove, 1992.
17. W. N. Venables and B. D. Ripley. *Modern Applied Statistics with S-Plus.* Springer, Berlin, 1994.
18. Statistical Sciences. *S-PLUS Guide to Statistical and Mathematical Analysis.* StatSci, MathSoft, Seattle, 1988,1993.
19. Alfio Marazzi. *Algorithms, Routines, and S Functions for Robust Statistics.* Wadsworth, Inc., Belmont, California, 1993.
20. Peter J. Rousseeuw. New applications of robust statistics. In *Compstat*, Barcelona, 1996.

Advanced Mathematical Tools in Metrology III
Edited by P. Ciarlini, M.G. Cox, F. Pavese & D. Richter

GAINING INFORMATION FROM BIOMAGNETC MEASUREMENTS

L. TRAHMS
Physikalisch-Technische Bundesanstalt Abbestr.2-12, D-10587 Berlin, Germany

Sensitive magnetic sensors called SQUIDs are the technical basis of biomagnetic measurements which allow insight into the function of human muscles and nerves. The analysis of biomagnetic measurement data is based on a multipole description of the current generator which is estimated using a best fit algorithm. The multipole approach can also be used for distinguishing flow and vortex current field components contributing to the measured signals. This is illustrated by experiments showing the presence of the two current field components in the repolarization phase of the heart beat.

1 Introduction

During the past twenty-five years, the advent of advanced information technology in biomedical measurement technique resulted in an enormous increase and improvement of diagnostic methods. Perhaps the most spectacular progress was achieved for computer supported imaging techniques, such as computer tomography (CT) or magnetic resonance imaging (MRI). These new methods provide insight into the anatomical structures of the human body which have become useful for basic research and medical diagnostics. In addition to structural information, the function of the human organs is of major importance for both researcher and clinician. This applies particularly to electrophysiologic processes in the heart and the brain that are associated with vital life functions.

A promising approach to obtain functional information on human tissue is based on ultrasensitive biomagnetic field measurements. Biomagnetic fields are generated by ionic currents that flow in muscles and nerves when they are active. Outside of the body, these fields are in the Pico- and Femto-Tesla range, i.e. more than one million times weaker than the Earth's magnetic field. Today, most sensitive magnetic field sensors, superconducting quantum interference devices (SQUIDs), are employed by cardiologists and neurologists for measuring the biomagnetic field of the heart and the brain, the magnetocardiogram (MCG) and the magnetencephalogram (MEG). These new diagnostic tools complement their conventional counterparts electrocardiogram (ECG) and electroencephalogram (EEG), and provide additional information that is inaccessible to other noninvasive techniques.

Since their invention in the late 1960es, SQUID technology has progressed enormously. From the early noisy single channel devices, low noise multichannel SQUID detectors have developed that record the biomagnetic activity around the brain or the heart. Today, some fifty large multichannel biomagnetic instruments are operating in hospital and research centers worldwide. For the coming years, the introduction of high temperature superconducting SQUIDs that can be cooled by cheap liquid nitrogen instead of expensive liquid helium will further ease the installation and operation of these instruments in clinics and propagate their use as a standard diagnostic tool.

In this article, section 2 shall give a brief view on the methodological background of the analysis of biomagnetic measurement data. Rather than trying to cover the wide range of different concepts, I will focus on two aspects of the biomagnetic approach, the information content of biomagnetic data and the inverse problem of localizing and characterizing biomagnetic generators. In section 3 some experimental background for these theoretical considerations will be given using medical measurement data. This representation refers to a study that was performed recently in a cooperation between the Physikalisch-Technische Bundesanstalt and the Rudolf-Virchow Medical Center [1].

2 Theoretical Background

The most ambitious aim of biomagnetic technology is the reconstruction of the current field, j, inside the human tissue at any instant, t, from the biomagnetic measurement data. This is an inverse problem. According to Hadamard`s definition, a well-posed inverse problem has a unique solution, which depends continuously on the measurement data. Usually, the solution of the biomagnetic inverse problem is defined as the current field that generates the smallest deviation (for instance in an L-2 norm) from the measured magnetic field data. However this solution is not unique, since there are many different ways to generate a given magnetic field outside of a volume conductor. In order to transform the ill-posed problem to a well-posed one, usually a physical model of the current is suggested that is characterized by a finite number of parameters. By approximating the parameters of the model generator to the biomagnetic measurement data, the solution of the well-posed inverse problem is found.

A very powerful class of models for biomagnetic generators is the multipole expansion of the current field, $\vec{j}(\vec{r})$ [2,3,4]. In terms of this expansion, the magnetic field $\vec{B}(\vec{r})$ is given by

$$\vec{B}(\vec{r}) = \sum_{n=1}^{\infty} \vec{B}_n(\vec{r}) + \sum_{n=1}^{\infty} \vec{B}_n{}'(\vec{r}) \tag{1}$$

with
$$\vec{B}_n(\vec{r}) = \sum_{m=0}^{n} \vec{B}_{nm}(\vec{r}, a_{nm}, b_{nm}), \tag{2}$$

where the electric multipole coefficients a_{nm}, b_{nm}

$$a_{nm} = g_{nm} \int_V \vec{j}(\vec{r}) \mathrm{grad}\left[r^n P_n^m(\cos\vartheta)\cos m\varphi\right] dv \tag{3}$$

$$b_{nm} = g_{nm} \int_V \vec{j}(\vec{r}) \operatorname{grad}\left[r^n P_n^m(\cos\vartheta)\sin m\varphi\right] dv \tag{4}$$

reflect current fields that have sources and sinks, i.e. the „flow" current fields,

and $$\vec{B}_n'(\vec{r}) = \sum_{m=0}^{n} \vec{B}_{nm}'(\vec{r}, \alpha_{nm}, \beta_{nm}) \tag{5}$$

containing the magnetic multipole coefficients α_{nm}, β_{nm}

$$\alpha_{nm} = \frac{g_{nm}}{n+1} \int_V \left[\vec{r} \times \vec{j}(\vec{r})\right] \operatorname{grad}\left[r^n P_n^m(\cos\vartheta)\cos m\varphi\right] dv \tag{6}$$

$$\beta_{nm} = \frac{g_{nm}}{n+1} \int_V \left[\vec{r} \times \vec{j}(\vec{r})\right] \operatorname{grad}\left[r^n P_n^m(\cos\vartheta)\sin m\varphi\right] dv \tag{7}$$

which reflect closed current fields, or „vortex" currents. In eqs. (4)(5)(6),(7),

$$g_{nm} = (2-\delta_{m0})(n-m)!/(n+m)!$$

and

$$\delta_{nm} = 1, \text{ if } m=0$$
$$\delta_{nm} = 0, \text{ if } m\neq0.$$

This decomposition of the biomagnetic field is related to the theorem of Helmholtz, which states that any vector field can be decomposed into its vortex and it flow part [5]. When applied to the current field, this means that

$$\vec{j}(\vec{r}) = \vec{j}_F(\vec{r}) + \vec{j}_V(\vec{r}) \tag{8}$$

with

$$\nabla \times \vec{j}_F(\vec{r}) = 0 \tag{9}$$

representing the properties of the vortex-free flow current, and

$$\nabla \cdot \vec{j}_V(\vec{r}) = 0 \tag{10}$$

representing the properties of the source- and sink-free vortex current. The two sums in eq. (1) reflect Helmholtz' decomposition of the current field in eq. (8) in terms of its action, the biomagnetic field.

Usually, not more than the first two or three terms of the expansion are employed for the analysis of biomagnetic measurement data. For the flow part of the field, the first term of expansion (2) is the current dipole

$$\vec{p} = \left(a_{11}, b_{11}, a_{10}\right). \tag{11}$$

It reflects the integral intensity of the flow current and generates the biomagnetic field

$$\vec{B}_1(\vec{r}) = \frac{\mu_0}{4\pi} \frac{\vec{p} \times \vec{r}}{r^3}. \tag{12}$$

The next terms of expansion (2) are the current quadrupole, which reflects the inhomogeneity of the current distribution, and the current octupole, which describes the asymmetry of the spatial extent of the current field [6].

For the vortex current, usually not more than its first component in expansion (5), the magnetic dipole is considered:

$$\vec{m} = \left(\alpha_{11}, \beta_{11}, \alpha_{10}\right). \tag{13}$$

It generates the magnetic field

$$\vec{B}_1{}'(\vec{r}) = \frac{\mu_0}{4\pi}\left(\frac{3(\vec{m}\vec{r})\vec{r}}{\vec{r}^5} - \frac{\vec{m}}{r^3}\right). \tag{14}$$

Note that this part of the current is seen by the magnetic field alone. In the corresponding analysis of the electric potential measured at the surface of the body, the contribution of the magnetic dipole source vanishes, like the contribution of any other vortex source of higher order[7]. This describes a fundamental difference between the electric and the magnetic measurement technique of electrophysiologic phenomena. It is expected from SQUID instrumentation, that this sophisticated measurement technique takes advantage of this benefit of the MCG and MEG. In the following experimental section, an example will be given that the additional information content of the biomagnetic field is not only of academic interest, but also physiologically meaningful.

On the basis of the first terms of the multipole expansion, the inverse problem can be solved. The origin of the multipole is estimated by finding the minimum of

$$\sum_{i=1,s}\left[\vec{B}_N(\vec{r}_i, a_{nm}, b_{nm}) + \vec{B}'_{N'}(\vec{r}_i, \alpha_{nm}, \beta_{nm}) - \vec{B}_e(\vec{r}_i)\right]^2$$

where $\vec{B}_e(\vec{r}_i)$ is the measurement value for sensor i, and $\vec{B}_N$ and $\vec{B}'_{N'}$ are the field values calculated for a multipole truncated down to order N or N', having a relative distance vector $\vec{r}_i$ to that sensor. For finding the solution of this nonlinear problem reliably, the Gauss-Seidel algorithm and the Levenberg-Marquardt algorithm are known to be most effective best fit algorithms. The best fit yields information on the location of the current and on its structure. As a matter of course, the amount of information depends on the number, s, of measurement data available.

3 Experimental data

Perhaps the best known example of an electrophysiologic process is the activity of the beating heart. The conventional approach to measure this activity is the electrocardiogram as it is shown for a healthy volunteer in figure 1. This signal is characterized by the slight bump indicating the activity of the atrium, followed by the sharp peak due to the depolarization of the ventricles which become repolarized some 200 ms later generating a slower and less intensive wave. Upon stress, usually the heart rate increases from, say, 60 beats per minute (bpm) to 140 bpm, as it is shown in the right chart. For a healthy person, this intervention should not change the morphology of the time dependent ECG signal significantly, except for the compression of the time scale.

In the two charts of figure 2, a magnetocardiogram measured simultaneously with the upper two ECGs is shown. Note that the MCG shows a change in morphology upon an increased heart rate. The unchanged shape of the depolarization signal of the MCG is contrasted by a change in polarity of the MCG during repolarization at the increased heart rate. The corresponding ECG does not show any comparable pattern. These data indicate that during the repolarization phase of the heart beat, the MCG displays phenomena which are not observed in the ECG.

This was investigated in more detail in a multichannel study where the precordial field was recorded using 37 MCG channels in parallel to 32 electric leads which were placed on both sides of the thorax. A most illustrative view on such

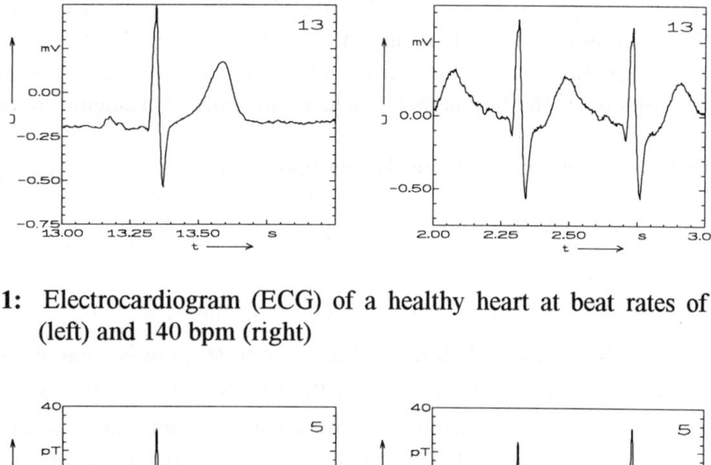

Figure 1: Electrocardiogram (ECG) of a healthy heart at beat rates of 60 bpm (left) and 140 bpm (right)

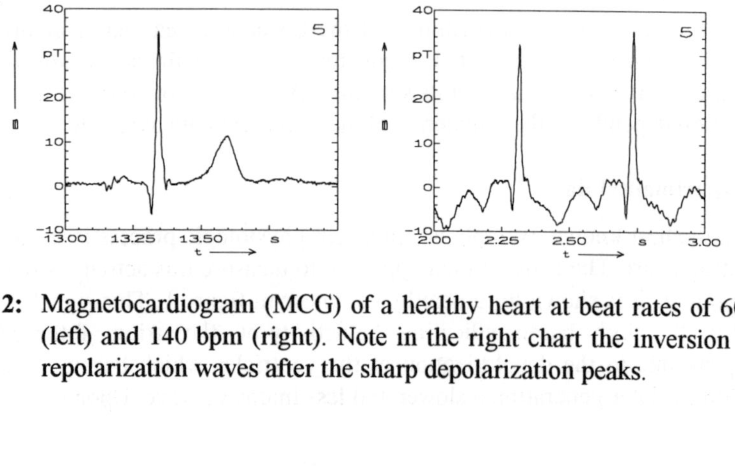

Figure 2: Magnetocardiogram (MCG) of a healthy heart at beat rates of 60 bpm (left) and 140 bpm (right). Note in the right chart the inversion of the repolarization waves after the sharp depolarization peaks.

multichannel data is the isocontour plot of the potential and field distribution at some time instant. Figure 3 depicts the situation for a time instant when the depolarization is active. The isopotential plots of these two charts show the potential distribution on a surface wrapped about the thorax with the left half representing the chest and the right half representing the back. The two isofield plots of figure 4 are a view on the precordial field plane parallel to the thorax. Apparently, both the ECG and the MCG pattern do not change significantly upon increasing the heart rate. Both patterns exhibit a dipolar structure having two

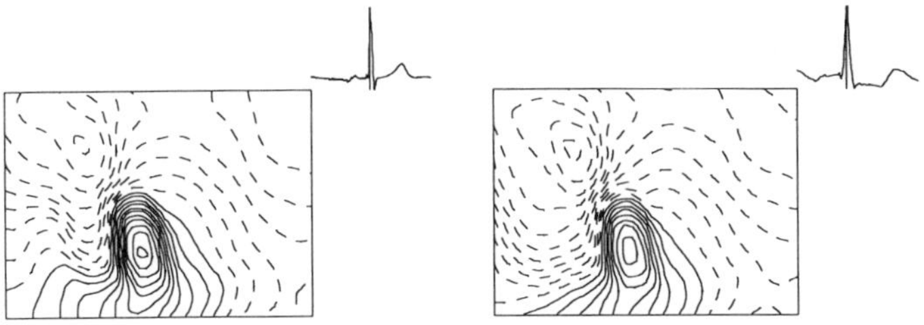

Figure 3: Isopotential plots of a 32-channel ECG of a healthy heart recorded on both sides of the thorax at beat rates of 60 bpm (left diagram) and 140 bpm (right diagram). The left (right) half of the two maps represents an area of 200 x 280 mm^2 of the chest (back); the solid (broken) lines represent negative (positive) field steps of 100 μV. Both charts show the **depolarization** phase as is indicated by the cursor position in the time signal at the upper corners of the two plots.

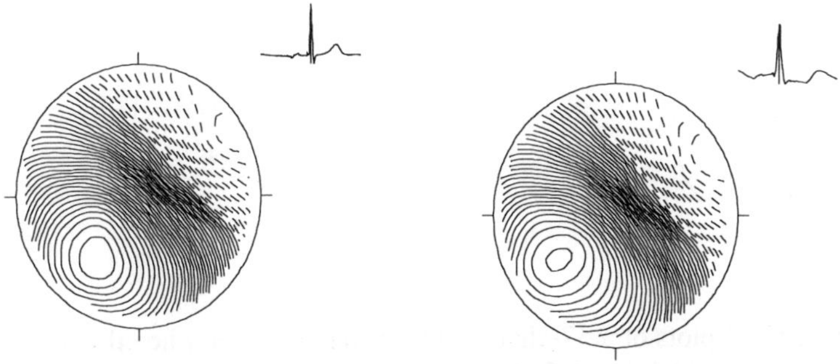

Figure 4: Isofield plots of a 37-channel precordial MCG of a healthy heart at beat rates of 60 bpm (left diagram) and 140 bpm (right diagram) for the **depolarization** phase. The area covered has a diameter of 210 mm, the solid (broken) lines represent negative (positive) field steps of 1 pT.

extrema, which does change neither in polarity nor in intensity at higher rates. Potential and field structures like these can usually be understood in terms of the current dipole, using eqs. (11) and (12), which is closely correlated to the so-called electric heart vector.

A different situation is met for the repolarization phase of the heart cycle. Figures 5 and 6 summarize the different behavior of ECG and MCG patterns upon an increase of the heart rate for this phase. Again, the isopotential pattern does not change significantly. In contrast, the isofield pattern recorded at the same time

Figure 5: Isopotential plots of a 32-channel ECG of a healthy heart recorded on both sides of the thorax at beat rates of 60 bpm (left diagram) and 140 bpm (right diagram). Details are the same as in figure 3. Both charts show the **repolarization** phase as is indicated by the cursor position in the time signal at the upper corners of the two plots.

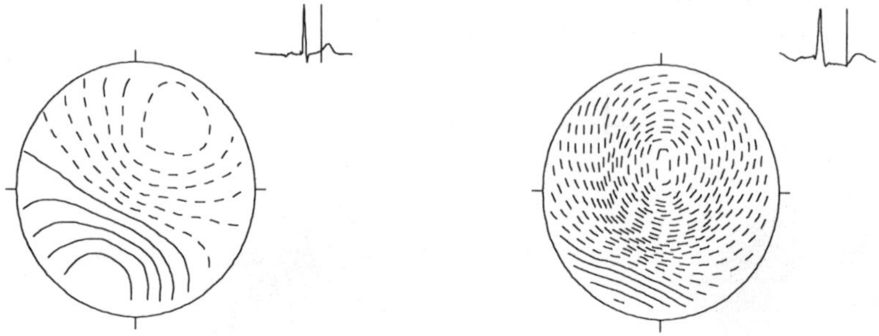

Figure 6: Isofield plots of a 37-channel precordial MCG of a healthy heart at beat rates of 60 bpm (left diagram) and 140 bpm (right diagram) for the **repolarization** phase. Details are the same as in figure 4. Note the change from a dipolar to a monopolar like structure of the pattern.

instant shows dramatic differences for the two heart rates.

This is a direct evidence of the existence of different sources of the two signals, ECG and MCG. It shows that there is a contribution to the biomagnetic field of the heart which does not affect the electric surface potential. Applying the considera tions of the previous section, this finding can be referred to the law of Helmholtz, and the observed changes in the MCG can be attributed to the vortex current field. In first order, the measurement data can be interpreted using the first two terms of the multipole expansion, i.e. the electric current dipole (eqs. 11 and 12) and the magnetic dipole (eqs. 13 and 14). By fitting the components of the two dipoles for some average location using the minimizing procedure described above, the results of the bar chart in figure 7 were obtained. It shows that both flow currents in terms of p_x and p_y, and vortex currents in terms of m_x, m_y, and m_z are generating the magnetic field of the heart in the repolarization phase. At an increased rate, only the magnetic dipole of the vortex contribution of the current field changes its direction and induces a change in the MCG. According to this interpretation, the vortex contribution is present in the current field during repolarization both at lower and at higher heart rates, but is the fact that it changes in front of a constant background of flow currents that reveals its presence unambiguously.

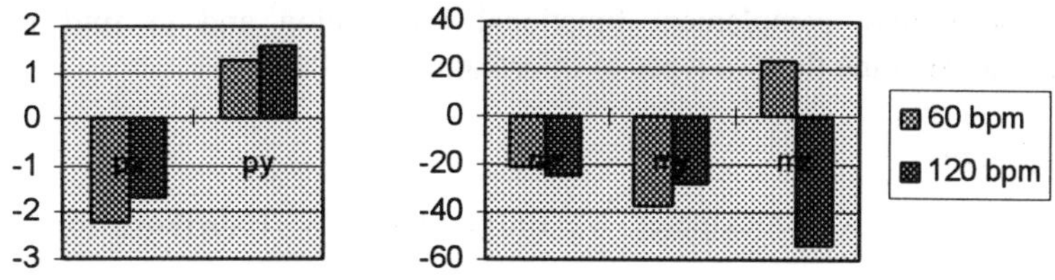

Figure 7: Bar charts illustrating the change of the electric current dipole moments p_x and p_y (left chart, measured in μAm) and the magnetic dipole moments m_x, m_y, and m_z (right chart, measured in nAm^2) upon changing the heart rate from 60 bpm to 140 bpm. Error margins are in the range of ±0.2 μAm for p and ±2 nAm^2 for m. Note the significant change in m_z, indicating a change in the orientation of the magnetic dipole.

4 Conclusion

Biomagnetic measurement technique based on SQUID sensors allows a new access to electrophysiologic processes. The method is superior to established measurement techniques, such as functional MRI, SPECT or PET, in that it measures a direct action of the electrophysiologic current with a temporal resolution of milliseconds. It is complementary to measuring the electric potential, since it provides more information of the current field. Fitting origin and components of a multipole to the measurement data yields diagnostically relevant information on the location and the structure of the current generator. However, this is still far from a functional image of the current field. On the basis of biomagnetic signals alone, a reconstruction of the current field with a satisfactory spatial resolution will be difficult to obtain, even if the measurement technique continues to improve in the future.

A breakthrough in biomagnetic technology might come from the field of information technology. A promising new approach is the addition of structural information to the functional biomagnetic data. For a more detailed view on distributed electrophysiologic activity in the tissue, the knowledge of the conductivity of the inactive tissue is useful. Its structure can be determined by a 3-dimensional MRI image of the region of interest. From these data, the boundaries of compartments with homogeneous conductivity are calculated by specialized image processing methods. As a result, a volume conductor model is added to the generator model, leading to a more realistic description of the physical situation. At the present stage, the addition of structural to the functional measurement data requires considerable effort in manpower and computer power, but it appears to be the most promising approach to obtain a detailed functional image of the human nerve or muscle tissue. It is a challenge to information technology to explore to what extent structural images may complement functional information and to find most efficient ways to combine the two types of information.

Acknowledgments

I would like to thank K. Brockmeier, M. Burghoff and W. Haberkorn for their assistance and advice.

References

1. K. Brockmeier et al, *J. Cardiovasc. Electrophysiol.* (1997) in press.
2. R. Plonsey and D.J. Fleming, *Bioelectric Phenomena*, McGraw Hill, (New York, 1967).
3. P.J. Karp et al, *Circulation Research* **47 1**, 117 (1980).
4. J.P. Wikswo and K.R. Swinney, *J Appl Phys* **56 11**, 3039 (1984).
5. J.A. Stratton, *Electromagnetic Theory*, McGraw Hill (New York, 1941).
6. M. Burghoff et al, in *Biomagnetic localization and 3d modelling*, eds. J. Nenonen et al. (Otaniemi, 1991).
7. W. Haberkorn et al, *Biomed. Technik* **37 S 2**, 156 (1992).

Advanced Mathematical Tools in Metrology III
Edited by P. Ciarlini, M.G. Cox, F. Pavese & D. Richter

INCREASE OF INFORMATION IN THE COURSE OF MEASUREMENT

PETER BÖLÖNI

National Office of Measures

P.O. 919, H-1539 Budapest, Hungary

Results of measurements are messages about the value of the measurand. As measurands can be described as discrete sources of information Shannon's formula is applicable to describe the missing quantity of information both before and after the measurement. Quantitative description of the measurement uncertainty provides the chance to use the discrete version of the Shannon's formula. By fixing the zero level of the necessary information to get the measurand completely known, different measurement results become comparable.

1. Indroduction

Measurements are important primary sources of quantitative information. Up to now the traditional definition of measurement emphasises its nature of process: Measurement is the set of operations having the object to determine the value of a quantity [1]. In describing the performances of measuring instruments used for measuring quantities of more complex nature like spectra, distributions or integrated values over a period, the definition of measurement by the theory of sets has some advantages: Measurement provides a transformation from the set of the possible values of the measurand to the set of possible results of measurements. The growing need for a tool to compare measurements of different nature and the common use of computers to control measurements and to process results call for a third definition related to the generally accepted information obtaining character of measurements. This third definition might be as follows: Measurement is a process decreasing the insufficiency of information about the measurand. The aim of this paper is to show a way to characterize the quantity of information obtained in the course of measurements.

2. Information and Quantity of Information

Information in a general sense is a certain concrete part of human knowledge, which can be described with words and phrases. From a scientific point of view information is a basic term of the theory of information and cannot be defined within the frame of this theory itself (just as the term "point" which cannot be defined in geometry and the term "time" which cannot be defined within the the physics). It is common to describe the method to measure the basic quantities of a science rather than to define the basic terms themselves, e.g. time is the quantity measured with a clock in specified manner. Information theory has a similar structure: instead of defining the term "information", a method is given to measure the quantity of information. As all complex information can be expressed with a set of the elementary information (quantum of information equals the information content of an answer of YES or NO of equal probability to a question to be decided), the quantity of information of any complex message can be expressed as a function of information content of the answers YES and NO comlpletely determining the message. Information content of a message is a function of the number of questions to be answered with YES and NO (but the quantity of information is not equal to the number of questions itself).

The occurrence of one particular event in a complete set with N number of elementary events can be identified with $\log_2 N$ number of questions if the strategy of searching is optimal. If the the figures of 0, 1, 2, 3, ... 9 can occur on the second digit of a digital display of a frequency meter with equal probability, then $n = \log_2 10 = 3.32$ questions (to be decided or answered with YES-s and NO-s) are usually necessary to identify the figure really occurring on the second digit. The amount of information provided by the second digit equals the information content of 3.32 answers of YES and / or NO. As the possible number of events is not a whole power of 2, in most cases $\log_2 N$ is not a whole number. (But the average number of necessary questions is still characteristic to the system of events or to the message or to any other source of information, just like the average number of children in families belonging to specified social groups is not a whole number although the number of children is always whole in real families.)

Probabilities of occurrence of elementary events are not equal in most practical cases. In English there are more words beginning with *L* than with *K*, so the probability of a single word message starting with *L* is higher than that of starting with the letter *K*. Probabilities of (elementary) events are not equal in this example. A system of events with different probabilities of occurrence changes the optimal strategy of search. If, for example, the *P (B)* probability of occurrence of the event marked with *B* is higher than 0.5 in a set of events comprising four events of *A, B, C* and *D*, then it is advisable to ask for the occurrence of event *B* first. If the answer is YES, the search is over. The average number of questions necessary to identify the i-th event equals $\log_2 \frac{1}{P_i}$ now denoting the probability of occurrence of the i-th event by P_i. This probability-based optimal strategy has a disadvantage of asking late for the less probable events. Identification of less probable events requires more questions in this way. The expected number of questions changes from the case of equal probabilities

$$I = \log_2 N \tag{1}$$

to the case of different probabilities

$$I = \log_2 \frac{1}{P_i} \tag{2}$$

where I is the expected or average number of necessary questions,
N is the number of independent and elementary events,
P_i is the probability of occurrence of the i-th event.

The unit of necessary questions seems to be identical with the unit of information which is 1 bit, the quantity of information of an answer to a question to be decided. (Most of the information is located in the question and not in the answer in this type of questions and answers in the common parlance.)

As a lot of messages have a random nature it is convenient to characterize the messages from the same source by the average quantity of information. Events of higher probability of occurrence contribute the average with a smaller number of necessary questions but with a higher frequency whereas events of less probability of occurrence contribute to the average with higher number of necessary questions but with lower frequency of occurrence. The number of necessary questions is weighted with the respesctive P_i probabilities this way, in the course of calculating the number of the average necessary questions or information. On average, the necessary information equals

$$H = \sum_{i=1}^{N} P_i \cdot \log_2 \frac{1}{P_i} \text{ bit.} \tag{3}$$

Formula (3) has been introduced by dr Claude Elwood Shannon. (It should be mentioned here that the author couldn't demonstrate the convergence of H characterizing a discrete source to a continuously variable source when the discriminations of possible events converge to zero. Since some model calculations have shown definite divergence, it is hard to compare the discrete and the continuous formulas for calculation the information obtained in the course of measurement.)

3. Prior Information about the Measurand

Both the increase of resolution of a measuring instrument and the accuracy of results are limited by the instability the of measurand and the ambient conditions, as well as the noise, the costs, and time consumption of the measurements. For this reason the results of measurements always have a limited number of digits. As measurement results are messages of series of decimal digits with limited lengths (completed with decimal points, units signs, etc.), the value of the measurand can be always regarded as an element from a finite set of possible values. Consequently, the possible values of the measurand can be described as a complete set of discrete values with P_i probability of occurrence of the i-th value. By specifying the respective probabilities for each possible value of the measurand, the probability distribution of the measurand is achieved prior to the measurement. Description of the measurand by a probability distribution prior to the measurements is unusual in metrology now but this is a way (if not the only way) to establish correspondance between the measurand and any other source of information dealt with the information theory.

Let us denote the resolution of the measuring instrument or the smallest detectable difference of the measurand with Δx , and the limits or range within which the actual value of measurand x lies with $L \leq x \leq K$. In principle the probabilities $P(x_i)$ that the actual value of the measurand fall in the range of

$$\left[x_i - \frac{\Delta x}{2}; \qquad x_i + \frac{\Delta x}{2}\right]$$

can be determined for all the x_i values. This set of events is complete because the actual value of the measurand can fall in one of these intervals only. Two more intervals can be added, if necessary, one to describe the cases when the actual value of the measurand is smaller than the L lower limit of the measuring range of the instrument and one if the actual value is higher than the K upper limit. But this slight refinement does not affect the principle. The actual type and parameters of $P(x_i)$, the prior probability distribution describes our information about the measurand before measurements. Substituting the respective $P(x_i)$ and Δx values in formula (3) one can calculate the H (0) information prior to the measurements. In fact, this quantity of information ought to be obtained in order to get the actual value of the measurand perfectly determined or known. This quantity of information is missing before the measurement.

Fig. 1. Complete set of events and minimum level of information in clinical thermometry

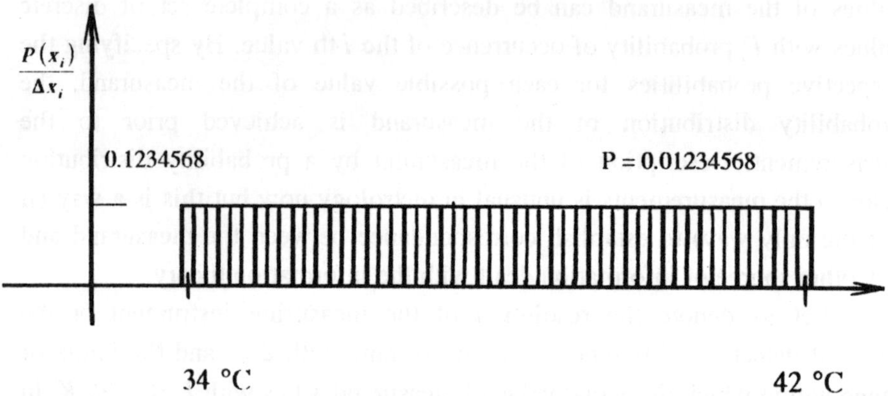

Fig. 1. demonstrates minimum level of information about a measurand with a mercury in glass medical thermometer as an example. Let the resolution of a medical thermometer Δx correspond to the generally expected uncertainty of 0.1 °C, the lower limit of the measurements range L = 34 °C and the upper limit K = 42 °C. Number of the possible events is

$$\frac{K - L}{\Delta x} + 1 = \frac{42 - 34}{0.1} + 1 = 81$$

When we know that the actual temperature has to fall in one of these intervals but we do not know more, then all the 81 intervals shall be assigned with the same probability of $P(i)$. As the actual temperature will fall in one of the intervals with $P = 1$ probability, sum of the 81 equal $P(i)$ probabilities equals 1. Thus, initially the value of $P(i)$ is 1/81 = 0.01234568 now. Using formula (3) one can calculate the information necessary to identify the actual temperature:

$$H(0) = 81 \cdot \frac{1}{81} \cdot \log_2 \frac{1}{\frac{1}{81}} = 6.34 \text{ bit.}$$

That is to say fixing the resolution and the range, 6.34 bit of information ought to be obtained to get the measurand perfectly described in this worst case (when nothing more can be supposed or exepted about the measurand but the resolution and the range). The index 0 indicates the state of minimum level of information.

Fig. 2. Level of information before measurements

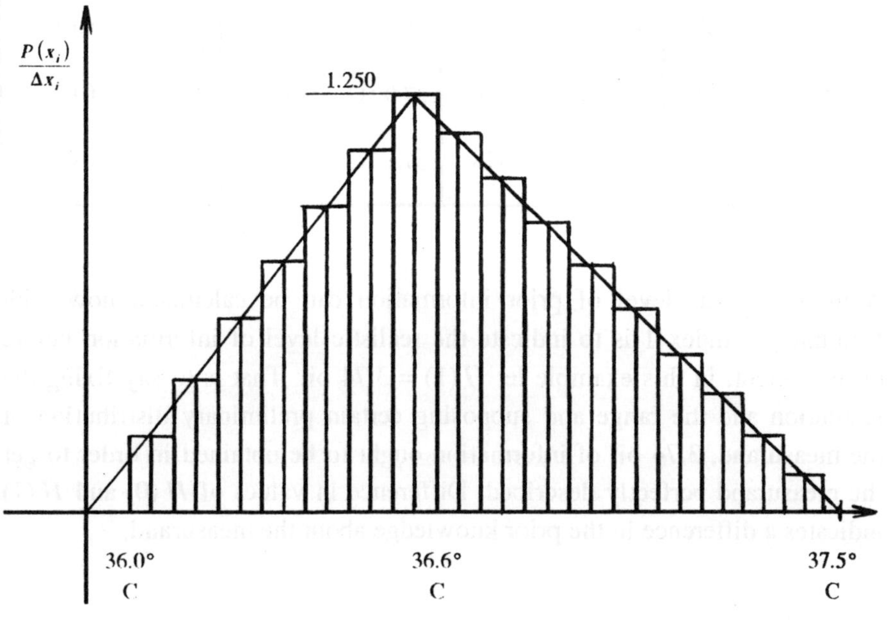

There is usually more information about the measurand than the resolution and the range before the measurements. The expectation of the observer excludes some part of measurement range and gives different probabilities for the rest between the maximum and zero. For a person looking healthy, temperatures less than 36 °C and more than 37.5 °C are not realistic. (These values correspond to the practice of medical thermometry used in Central Europe.) In this limited range the mode is about 36.6 °C. Let us suppose in this example that probabilities from this maximum value decrease to zero on both sides of the mode with constant slope according to the observers' opinion. Table 1 lists the respective $P(x_i)$ probabilities.

Table 1

36.0 °C	0.0096	36.4 °C	0.0865	36.8 °C	0.0987	37.2 °C	0.0460
36.1 °C	0.0289	36.5 °C	0.1058	36.9 °C	0.0855	37.3 °C	0.0329
36.2 °C	0.0481	36.6 °C	0.1250	37.0 °C	0.0724	37.4 °C	0.0197
36.3 °C	0.0673	36.7 °C	0.1118	37.1 °C	0.0592	37.5 °C	0.0066

A more realistic level of prior information can be calculated now with formula (3). Index 1 is to indicate the realistic level of information before measurement. In this example let $H(1) = 3.74$ bit. That is to say fixing the resolution and the range and supposing certain preliminary distribution of the measurand, 3.74 bit of information ought to be obtained in order to get the measurand perfectly described. Difference in values of $H(0)$ and $H(1)$ indicates a difference in the prior knowledge about the measurand.

4. Uncertainty of Measurement Results and Residual Insufficiency of Information

Apart from the specification of the measurand and the conditions measurement results consist of

- the value for estimation of the true value of the measurand,
- the uncertainty of the measurement result and
- perhaps the specification of the supposed type of probability density function of results (or errors).

Supposing a type of distribution function of observations (or errors) and using the parameters characterizing the uncertainty of results, one can calculate the $Q(x_i)$ probabilities for the value of measurand falling in the range of

$$\left[x_i - \frac{\Delta x}{2};\quad x_i + \frac{\Delta x}{2}\right].$$

In most part of the measuring range Q will not differ practically from zero. For the remaining few x_i values the $Q(x_i)$ probabilities differ from zero and can be substituted in formula (3)

$$H(2) = \sum_{i=1}^{N} Q(x_i) \log_2 \frac{1}{Q(x_i)} \text{ bit.} \tag{4}$$

Index 2 is to indicate the quantity of missing information characterizing the level of information about the measurand after the measurement is completed.

Let us suppose the result to be (36.8 ± 0.1) °C. As this x_m result is the best possible estimation of the value of the measurand, the probability of $Q(x_m)$ is the highest, say 0.96. The uncertainty specification of 0.1 °C allows the measurand to be 36.7 °C and 36.9 °C as well. The supposed distribution function of results permits one to calculate the probabilities of all particular results (not only the maximal one). Let both the 36.7 °C and 36.9 °C results have Q = 0.02 probability and the rest of the x values practically zero. Shannon's formula (3) allows again to calculate the missing part of information. It is

$$H(2) = 2 \cdot 0.02 \cdot \log_2 \frac{1}{0.02} + 0.96 \cdot \log_2 \frac{1}{0.96} = 0.28 \text{ bit.}$$

Fig. 3. Level of information after measurements

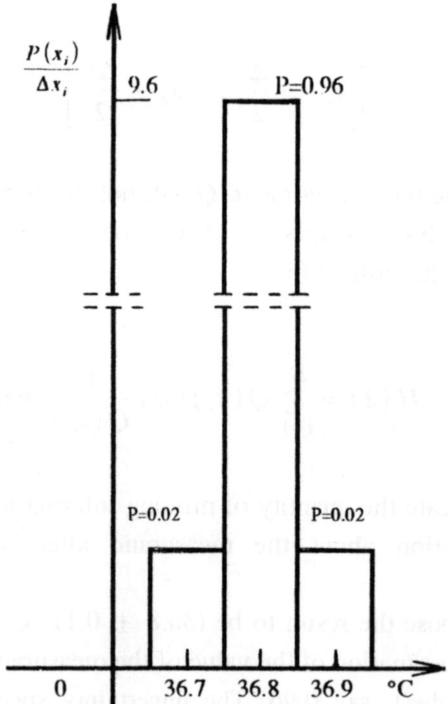

A much better level of information has been calculated now from the probability distribution of results (and not from the possible values of measurand) with formula (4). In this example $H(2) = 0.28$ bit. That is to say having the result and supposing some distribution of it, 0.25 bit information ought to be obtained to get the measurand perfectly described. The difference between the values of $H(0)$ or $H(1)$ and that of $H(2)$ indicates the difference between the prior knowledge and the amount of knowledge after the measurement.

It might be interesting to note that if the result were associated with an uncertainty less than $\Delta x/2$, then the only $Q(x_m)$ probability would be equal to 1 (see *Fig. 4.*). In this case Shannon's formula gives $H(p) = 0$, that is to say no information is needed to complete our knowledge about the measurand if the result does not have an uncertainty more than $\Delta x/2$.

$$H(P) = P \cdot \log \frac{1}{p} = 1 \cdot \log \frac{1}{1} = 0 \text{ bit.} \qquad (5)$$

This is an indication of the principle perhaps expressed by Hinchin for the first time: Shannon's formula measures the quantity of information necessary to eliminate the insufficiency of information about the state of a source rather than measuring the information content of the source itself [4]

Fig. 4.

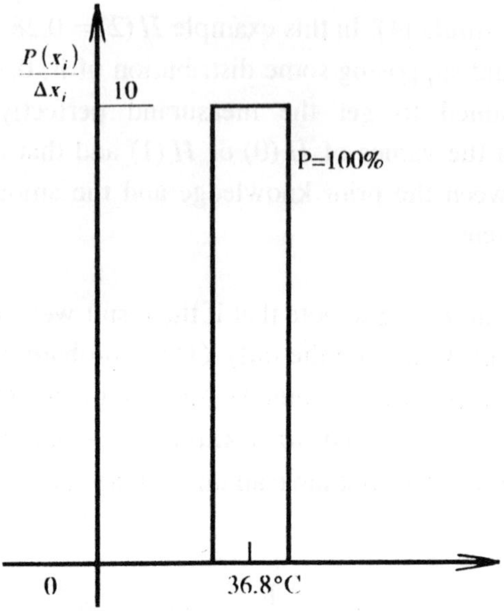

5. Quantity of Information Obtained in the Course of Measurement

In order to determine the information obtained by a measurement the previous considerations shall be summed up and completed first:

H (0) or *H* (1) amount of information is needed to eliminate the insufficiency of information prior to the measurement.

H (2) amount of information is needed to eliminate the remaining insufficiency of information after the measurement.

$H(p) = 0$, no information is needed when the measurand (or the state of information source) is perfectly determined.

Mathematical properties of equivalency and additivity are valid for two quantities of information both of them described by Shannon's formula. Let us denote the level (or amount) of information associated with the measurand prior to the measurement by *G* (0) or *G* (1) and by *G* (2) after the measurement. Adding H (0) or H (1) information to the prior value of $G(0)$ or $G(1)$, respectively, one comes to the state of perfectly determined $H(p)$ when no more information is needed. Similarly, adding $H(2)$ information to the $G(2)$ level of information after the measurement, one comes to the state of perfectly determined $H(p)$ when no more information is needed.

$$G(0) + H(0) = H(p) \text{ or}$$
$$G(1) + H(1) = H(p) \text{ or}$$
$$G(2) + H(2) = H(p). \tag{6}$$

As $H(p) = 0$ according to (5) and the H values calculated with the Shannon's formula are not negative, we have got negative values for all the levels of information characterizing the measurand (before as well as after the measurement):

$$G(0) = -H(0) \text{ or } G(1) = -H(1) \text{ and } G(2) = -H(2) \tag{7}$$

The $H(m)$ amount of information obtained by the measurement can be calculated finally as a difference of levels of information after and before the measurement:

$$H(m) = G(2) - G(0) \quad \text{or} \quad H(m) = G(2) - G(1) \tag{8}$$

Substituting the calculable values from (7) we get:

$$H(m) = H(0) - H(2) \quad \text{or} \quad H(m) = H(1) - H(2) \tag{9}$$

Supposing equal probability of all temperatures in the range, the measurement resulted in $H(m) = 6.34 - 0.28 = 6.06$ bit of information. Having some experience in medical thermometry, that is using some previous information, $H(m) = 3.74 - 0.28 = 3.49$ bit information is gained. As the information quantity of any message about the state of source of information or any result of measurement calculated with Shannon's formula is always more than zero, the level of information or quantity of information of the source or the measurand has to be less than zero (to meet the requirement of formula (5).

6. Example

Let us have two pressure gauges of Bourdon type with 0 ... 100 bar measurement range each, with resolutions 0.5 bar and 0.1 bar and the error limits specified for gauges 1 bar and 0.1 bar respectively. How much more information can the second gauge supply about the measurand? (For sake of simplicity we suppose that the measurand does not have any preferred range or value and the probabilities of all the results are equal within the range of the error limits).

Number of the different possible values of measurand for the first gauge is 200 and 1000 for the second. The number of permissible results is 5 with the first gauge as apart from the result there are 1 bar/0.5 bar = 2 possible values within the error range in both sides of the result itself . For the second gauge the number of permissible results is 3. The difference of supplied information $H(D)$ equals:

$$H(D) = [\, H(0)_2 - H(2)_2 \,] - [H(0)_1 - H(2)_1 \,],$$

$$H(D) = [\, \log_2 1000 - \log_2 3] - [\log_2 200 - \log_2 5 \,],$$

$$H(D) = 3.06 \text{ bit}.$$

more information is provided using the second gauge.

7. Doubts and Further Considerations

As it was mentioned earlier, no bridge has been found up to now between the continuous and the discrete version of the theory of information for the interpretation of measurement results. The problem lies in the fact that most of the metrologists use continuous functions for modeling because of the simplicity of use and the extensive practice of application. But the description of the measurand with continuous values supposes exact results i.e. results without uncertainty. Such a result could only be written as an infinite long message which ought to contain infinite amount of information. The continuous version of the theory gives similar answers to some questions but the results were not found numerically convertible.

It is easy to identify the source of information and the measurand at a glance. But some principal difficulties must be overcome to be able to use the above described approach for metrological purposes. The true value of the measurand is usually not known whereas the message is considered to be completely known (before coding). Measurements often change the value of the measurand whereas it is hard to find any case when coding changes the original state of the source. Noise does not affect the coding but affects the measuring processes, and measurement results are messages usually not intended to be decoded at all. But the chance to characterize the information quantity of measurements gives good reason for efforts of further improvement of the application of information theory in metrology.

7. Summary

Determining the prior distribution probabilities of the measurand's different values, Shannon's formula becomes applicable to calculate the missing quantity of information about the measurand. As all the measurement results are uncertain to certain extent, some insufficiency of information will remain after each measurement. Fixing a zero level, the quantity of information provided by different measurements can be compared.

References

1. BIPM; IEC; IFCC; ISO; IUPAC; IUPAP; OIML: International Vocabulary of Basic and General Terms in Metrology 2nd Edition, ISO 1993.
2. Periodica Polytechnica Ser. Mech. Eng. Vol. 37, No 4, PP 223-235, 1993.
3. Claude E Shannon - Warren Weawer: The Mathematical Theory of Communication, University of Illinois, 1972.
4. Ja. Hincsin: Ponjatie entropii v teorii verojatnoszyej. Uszpehi matematicseszkih nauk, 1953.

Advanced Mathematical Tools in Metrology III
Edited by P. Ciarlini, M.G. Cox, F. Pavese & D. Richter

A FRAMEWORK FOR MODEL VALIDATION AND SOFTWARE TESTING IN REGRESSION

B BUTLER
Scientific Software Section, NPL, Teddington, TW11 0LW, UK

Regression objectives are considered from the viewpoint of the metrologist, leading to an analysis of the areas in which regression results are unsatisfactory. Key factors are identified and a set of tests is proposed to support model selection and validation (by the metrologist) and software testing (by the software developer).

1 Introduction

The motivation and key concepts are introduced in this section. In Section 2, the pitfalls which can cause regression to fail are identified. Key factors are noted in Section 3. The resulting criteria for successful regression are mentioned in Section 4. A procedure for model selection and validation is provided in Section 5, and some software testing methods summarised in Section 6. Conclusions are drawn in Section 7.

Valid metrological results must be traceable to standards. Software typically forms part of the traceability chain. Tests exist for the analysis methods and software, but there is often little guidance as to *which* tests to use and *how* to use them effectively. This is true even though the software tools are (arguably) becoming easier to use, and particularly now that many appear in non-specialist packages such as spreadsheets.

The growth in certification schemes and Quality Management Systems have accelerated moves towards *formal*, *objective* quality requirements, rather than expert, though *subjective*, opinion. Therefore, it is timely to consider the topic. This paper introduces a procedure for regression model validation and does not provide an analysis, comparative or otherwise, of methods and tests. The latter is a much wider topic and would probably be application-specific.

The motivating activities were (i), development of a regression package for calibration curves at NPL; (ii), analysis of platinum resistance thermometer (PRT) data[1], and (iii), validating the CMM Parametric Error Model (PEM). The PRT data was assembled from different sources, and a functional representation was required for comparison with the ITS-90 scale and to be used as a basis for fitting with Debye functions. The PEM[2] is a semi-empirical model, in that 18 functions model the positional and rotational errors in a CMM, with the functions themselves having an empirical form (typically polynomials and splines).

Regression objectives must be chosen with *fitness for purpose* in mind if the user is to be satisfied. Hence, consideration is needed of (i), the reasons for failure, and (ii), the nature of the regression objectives.

Some terminology is required. (Regression) *model selection* is concerned with *which model* to use. It is part of a *design* procedure, and hence is *iterative* in nature. The *decision tree* has many options (e.g., model function form, nature and use of error models). (Regression) *model validation* is concerned with asking questions and providing *statistical* answers in support of model selection. Initially, high-level *between-family* decisions are required, which in turn motivate low-level *within-family* decisions. For instance, sequential decisions might be to use (i) an empirical model (ii), a polynomial (both high-level), (iii), least-squares norm and (iv) degree 5 (both low-level). Model selection is supported by tools, but is ultimately the user's responsibility.

Testing of (regression) software differs in that it comprises *mathematical* tests on the *implementation* of the selected model. The key aim is *generic correctness* of the regression tools, with regard to *reliability* (the software tools perform to specification), *robustness* (they perform adequately for extreme cases) and *efficiency* (they use resources well, particularly data precision).

2 Problem areas

Deficiencies can occur in respect of the *data*, both its *quantity* and *quality*. Quantity is governed both by the *number* of data elements and by its *distribution*. Circle fitting is a well-known example. At least three distinct points are required. If these points lie on a small arc, the traditional parametrisation is ill-conditioned and the individual parameters have a large uncertainty. Even if the points are (apparently) well-distributed, a finite sample will generally underestimate the form (systematic) error; in pathological circumstances, it could miss it completely! With regard to *quality*, the errors should be tighly-controlled and/or well characterised. The problem is reduced by better metrological awareness and obtaining more *representative* data, which adds to costs.

Another potentially weak point is *problem formulation*. Often, users have an intuitive understanding of the data and its errors, which needs to be incorporated in the regression. It often helps to prepare a formal *user requirement document*, emphasising objectives and constraints. This is supported and informed by documentation, model validation and, possibly, an intelligent user interface.

Algorithm design continues this theme, though the balance of control shifts towards the tools. Only low-level fine-tuning remains, which is heavily constrained by implementation details. This stage is the "last chance" for validation. Whether formal or not, a *functional specification* is the outcome.

The *software implementation* can fail because of the presence of (i) unstable methods; (ii) logic errors; (iii), inadequate exception handling; (iv), language errors due to the developer and/or the compiler. Many of these are subtle and can persist after gross errors have been removed. Numerical stability and correctness in some statistical procedures, including regression, have been considered[3].

3 Issues governing the quality of the regression

Poor *data* limits what can be achieved (see Section 2). Also, for multivariate data, some independent (control) variables might not be *statistically significant* in the regression.

When developing the error model, the following need to be considered: (a) which variables are subject to error; (b) the distribution of the (weighted) residuals; (c) correlation of the residuals; (d) analysis of systematic and stochastic components and (e) size. If more than one variable is subject to error, generalised distance regression (GDR)[4] should be used.

Firstly, the error covariance matrix V should be estimated; it is often necessary to use information from other sources. Let $X^{\mathrm{T}}V^{-1}X$ be the Gauss-Markov estimator of the parameters. If, as is often the case, the errors are uncorrelated, V is diagonal, so errors weighted by the inverse of the diagonal values are homoscedastic. If the errors are *unbiased*, the Gauss-Markov estimate will be unbiased[5]. However, if the errors have a non-Normal distribution, least-squares (l_2-norm) estimation should be replaced by another fitting criterion such as l_1 or l_∞. Clearly, errors can depart from Normality when there are outliers, in which case either a rejection[6] or accommodation (robust estimation[7,8] (see also the paper by W. Stahel in this volume)) strategy should be used.

Additional/prior information (e.g., from experience and handbooks) is not only useful for defining the error model[9]. Users often require the regression model to satisfy simple constraints, e.g., that strain never decreases as load is increased. This amounts to a *shape constraint* on the fit. More generally, users prefer simple fitting functions, which are much easier to interpret. This shape can be lost if the data is poor and the model functions are not *designed* to satisfy the constraints.

The *functional requirements* help to define the overall regression aims. Requirements include: (a) data reduction; (b) parameter estimation; (c) parameter identification/reduction; (d) evaluating estimable functions of the parameters, and (e) estimating uncertainties of parameters and derived quantities. In the case of (a), apart from evaluation, the actual model function is immaterial. With (b), the model is semi-physical at least, and there are often parameter constraints. In the case of (c), there is considerable (near-)redundancy in the parameters, and part of the task is to find a linear combination such that a lower-dimensional regression model is adequate (see paper by H-K Liu elsewhere in this volume). The singular value decomposition (SVD)[10] is a very useful tool for such problems. As to (d), estimable functions include the regression model function itself, and, e.g., for polynomials, its analytical derivative, etc. It is also usually necessary to estimate the uncertainties of computed parameters[9].

4 Properties of a good fit

For a given model, regression can be deemed successful if (a) the residual metric is consistent with the predicted error metric; (b) the residual metric is a minimum over all fits satisfying relevant constraints; (c) the residuals have the expected distribution of the errors; (d) the residuals have a symmetric distribution centred on zero; (e) the values/uncertainties of critical parameters are consistent with what was expected; (f) constraints on the fit and the parameters are satisfied; (g) when evaluated, the fit is "taut" and, if additional data is obtained, the deviations are consistent.

Note that properties (a) to (d) are based on the assumption that the residuals are unbiased and consistent estimates of the errors. Therefore their behaviour can be compared with the error model, each validating the other, as well as being used to find the "best" regression model. The user often has a target error measure, as in (a), or knowledge of the parameters, as in (e). Note that, in (b), successful regression involves trade-offs between objectives. Conditions (c) and (d) relate to residual distribution. The former can be tested using a t-test; the latter by using a binomial test on the signs of the residuals and a χ^2-test applied to the lengths of runs-up or runs-down. If (f) is not satisfied, the constraints should be built into the regression procedure. As for (g), a taut fit lacks wobbles between data points (wobbles are symptomatic of overfitting and/or an ill-conditioning regression problem). The deviations test is a special form of regression cross-validation, which can be generalised to subsets of the original data.

5 Example: model validation procedure for regression

The following procedure presents a practical *specific* example of the application of the concepts above in metrological regression. For ease of exposition, the validation tests[5,7] are described for fitting algebraic curves, but they can readily be extended to surfaces.

1. *Analyse an appropriate residual measure s.* Compare models by ranking their s values. Use a χ^2-test on each s based on s/σ, where σ is the predicted s. Perform a parameter significance F-test.

2. *Analyse the Pearson product-moment correlation coefficient.* Note that r^2 (coefficient of determination) is $\text{var}(\hat{y})/\text{var}(y)$ (where $\hat{y}$ is the fit to ordinates y). After applying the Fisher transformation to r, a z-test can be applied.

3. *Studentise the parameters* and perform a t-test to determine their significance.

4. *Plot residual measure* against regression dimension (number of regression parameters). Choose the dimensions at which the residual measure begins to saturate and/or reaches the predicted noise level of the data.

5. If the basis functions (observation matrix columns) are orthogonal (e.g., when fitting Fourier series to uniform data or Chebyshev-basis polynomials to data at the Chebyshev points), *plot successive regression parameters against regression dimension.* Analyse as in (4.).

6. *Plot the fit* at densely-spaced abscissae and look for "wobbles". If possible compute the arc-length and plot this against regression dimension. Choose the dimension at which the arc-length begins to "plateau". Check also that the shape is as expected.

7. *Plot the residuals* and inspect them for trends. If appropriate, test the distribution hypothesis (e.g., use Shapiro-Wilk or Kolmogorov-Smirnov tests as appropriate). Test for outliers (e.g., using Grubbs test if residuals are assumed to be Normal[6]).

8. *Plot the residuals against each regressor variable* and look for systematic behaviour, e.g., if the residuals appear to have an exponential dependence on the regressor variable, an exponential transformation of that variable might improve the fit.

This procedure can be expanded. It is designed to be *informative* and *practical,* since most regression software will provide information on which the tests can be based.

6 Software testing overview

Software testing aims to verify that software has *mathematical* correctness. This section outlines how it may be applied to regression. Note that packages such as spreadsheets and allied software blur the distinction between data-entry and "programming". Hence the need for software testing is not restricted to "traditional" developers.

Formal *comparative testing* is based on reference software and results[3,11]. The aim is to show that results obtained by software under test are *numerically consistent* with reference results; performance measures supporting this comparison have been considered[3]. Less formally, comparative testing is used during general software development to compare old and new versions; this *regression testing* (of software) is not to be confused with testing regression software!

Data generators extend the power of comparative testing. Indeed, data generation and regression are dual problems. Moreover, data generation is a forward problem, whereas regression is an inverse problem[12]. There is no need for reference software if the data generator has a forward error which is less than the backward error of the regression problem[13].

Internal/consistency testing is useful even when reference results are unavailable. (Mathematical) *invariance properties* of the solution are identified. In the case of geometric feature fitting: (a) the residuals are invariant with respect to rototranslations; (b) feature parameters are affected as predicted, e.g., the sphere centre is translated by that translation.

7 Conclusions

Concepts have been formalised and categorised as part of an overview of regression. This framework leads to a procedure for regression model validation. Both are being developed with feedback from their application in project work at NPL.

Research continues into integration of model selection and validation, i.e., making more use of validation information to improve the model. So far, an error model is assumed in parts of the procedure but perhaps this could be relaxed and robust methods, such as nonparametric statistics (see papers by P Ciarlini and P M Harris et al elsewhere in this volume), used instead. This would lead to fewer assumptions, but validation is still needed for model selection.

References

1. A.J. Abackerli, B. Butler, and M.G. Cox. Application of advanced mathematical and statistical tools to temperature-measurement data. Technical report, NPL, CMOT, 1997. In preparation.
2. P.C. Cresto. Self-calibration with application to CMM geometry error correction. In *Proceedings of the International Workshop: Advanced Mathematical Tools in Metrology*, pages 167–174, Singapore, 1994. World Scientific.
3. B.P. Butler, M.G. Cox, A.B. Forbes, S.A. Hannaby, P.M. Harris, and S.M. Hodson. *Statistics software qualification—Reference data sets*. Royal Society of Chemistry, Cambridge, 1996. edited by B P Butler, M G Cox, S L R Ellison, W A Hardcastle.
4. A. B. Forbes. Generalised regression problems in metrology. *Numerical Algorithms*, 5:523–533, 1993.
5. G. A. F. Seber. *Linear regression analysis*. John Wiley and Sons, New York, 1977.
6. V. Barnett and T. Lewis. *Outliers in Statistical Data*. John Wiley and Sons, Chichester, 1984. 2nd Edition.
7. NAG. G02 chapter introduction. In *NAG Fortran Library Document, Mark 17*. The Numerical Algorithms Group Ltd, Oxford, UK, 1996.
8. D.F. Vecchia and J.D. Splett. Outlier-resistant methods for estimation and model-fitting. In *Proceedings of the International Workshop: Advanced Mathematical Tools in Metrology*, pages 143–154, Singapore, 1994. World Scientific.
9. ISO. Guide to the expression of uncertainty in measurement, 1993. ISBN 92-67-10188-9, (1st edition).
10. G. H. Golub and C. F. Van Loan. *Matrix Computations*. J Hopkins Univ. Press, Baltimore, USA, 1989.
11. G.T Anthony, B.P Butler, M.G. Cox, A.B. Forbes, S.A. Hannaby, P.M. Harris, B. Bittner, R. Drieschner, R. Elligsen, H. Groß, and J. Kok. Chebyshev reference software for the evaluation of coordinate measuring machine data. Technical Report EUR 15304 EN, NPL, Teddington, UK, 1993. Published on behalf of the CEC.
12. B. Butler, M. Cox, A. Forbes, S. Hannaby, and P. Harris. A methodology for testing classes of approximation and optimisation software. In R.F. Boisvert, editor, *The Quality of Numerical software: Assessment and Enhancement*, London, 1997. Chapman and Hall. To appear.
13. A.B. Forbes. Validation of software for dimensional metrology. Technical Report DITC 225/94, NPL, Teddington, UK, 1994.

Advanced Mathematical Tools in Metrology III
Edited by P. Ciarlini, M.G. Cox, F. Pavese & D. Richter

CERTIFICATION OF ALGORITHMS FOR DETERMINATION OF SIGNAL EXTREME VALUES DURING MEASUREMENT

A.G.CHUNOVKINA, A.V.CHURSIN
D.I.Mendeleyev Institute for Metrology (VNIIM),
19, Moskovsky pr.,St.Petersburg, Russia

A common approach to the certification of algorithms for processing experimental data during measurements, as applicable to the algorithms used to determine the location and value of a valid signal extremum from the results of its measurement, is developed and rendered concrete. The results of metrological certification of concrete algorithms are presented. Their analysis is carried out, and possible applications are outlined.

1 Introduction

Specific character of application of mathematical methods in metrological practice is conditioned, first of all, by the existense of errors of initial data (both random and systematic) and the possible deviations of real data from the accepted mathematical models. Besides, asymptotic properties of obtained estimates are frequently investigated in mathematics, whereas a number of experimental data is always limited. From the aforesaid, it becomes evident, that, when using mathematical methods for processing measurement results, it is essential to investigate them additionally. When working out such an investigation , it is necessary to take into account that in metrological practice the information about the properties of experimental data are quite limited. That is why the user requires the information about the algorithm properties for a certain set of initial data models and different criteria of optimality rather than its own optimality for a certain criterion and initial data model.

A common approach to the metrological certification of algorithms for processing data in measurement, which has been suggested in [1, 2], makes it possible to solve the problem under consideration in the uniform metrological context. The idea of the metrological algorithm certification consists in their investigation using a set of typical models with the aim to define algorithms characteristics, the accuracy ones first. The final goal of the processing algorithm metrological certification is to provide an objective criterion for comparing the algorithms and to realize the rational choice of an algorithm in a particular measurement problem. It should be stressed that certification is carried out for a uniform group of algorithms intended for solving similar measuarement problems. The main characteristics of algorithms and the set of the typical models of initial data are specified, as applied to this group.

2 Statement of the Problem

The paper presents the results of the evolution and concrete definition of the common approach to the algorithm metrological certification as applicable to the algorithms for determining the location and value of valid signal extrema.

Let us represent the measurement signal as: $x(t) = s(t) + n(t) + \varepsilon(t)$, where s(t) - is the valid signal, n(t) - is the systematic noise (drift of the zero line), $\varepsilon(t)$ - is the random noise[3].

Processing of the measurement signal with the aim to determine the location and value of the extremum includes smoothing of the random noise, correction of the zero line drift and estimation of the parameters to be found. To solve the problem of smoothing the random noise, linear filters with impulse functions $h_1(\tau) = 1/T \exp\{-\tau/T\}$, $h_2(\tau)= \tau/T^2 \exp\{-\tau/T\}$, $h_3(\tau)=1/(1-\alpha T)\{1/T \exp\{-\tau/T\}-\alpha\exp\{-\alpha\tau\}\}$ are considered. The last filter considered accomplishes a partial correction of the zero line drift. To refine the estimates of the required parameters, we investigate the algorithms used for processing of a discrete sequence of initial data based on the local square approximation close to extremum (A), on the local cubic approximation close to extremum (B) as well as on the least-square method for parabola model (C).

3 Scheme of the Metrological Certification of Algorithms for Processing the Experimental Data

A scheme of the metrological certification of algorithms for determining the extremum location and value of the valid signal on the basis of its measurement, results includes the following steps:

a)Choice of initial data models. The models are established separately both for the valid signal and for the measurement errors. The essential requirement for the model sets to be formed is their relative simplicity and variability which allows to obtain the estimates of algorithm characteristics quite easily and to investigate the robustness of the algorithm in respect to the variation of the model. A list of models is not completed and it can be supplemented, reasoning from the specific metrological problem.

For s(t), the following models are proposed to consider as the principle ones:

- Gaussian model $s(t)=H \exp\{-4 \ln 2 (t-t^*)^2/ D^2\}$;
- Lorentz model $s(t)=H /\{1+4 (t-t^*)^2/ D^2\}$;
- various combinations of these two models, including those used for modelling of asymmetric peaks;

- common form model which is specified by the height of the peak (H), by its width (D), by the location of extremum (t*). This model is considered in the course of the given investigation.

For n(t) , the following models are proposed: $n_1(t)=0$, $n_2(t)=\delta_0+\delta_1 t+\delta_2 t^2$ and others. During this investigation both of the above-listed models are considered.

Lastly, for ε(t) it is proposed to use independent and identically distributed random quantities and random processes. The model of the first type is used in the certification of algorithms for processing of discrete sequences of measurement results. The models of the second type are used in the investigation of linear filtering algorithms.

b) Choice of the metrological characteristics of algorithms. Metrological characteristics of the algorithm are the characteristics of a result, obtained using the algorithm under investigation. They are to be selected as metrological ones.

The error of the algorithm consists of the following components:

- methodical one, caused by a specific realization of the algorithm in the absence of initial data errors. It is usually characterized by limits and correction,

- transformed one, caused by the transformation of initial data errors into a total error of the result in the course of processing. The transformed errors have systematic (for non-linear algorithms) and random components. The systematic component is commonly characterized by the limits and expectation, whereas the random one - by the standard deviation (an analogue of standard uncertainty) and by the confidence interval (an analogue of expanded uncertainty).

c)Evaluation of algorithm characteristics. There are two principal ways of evaluating the characteristics of algorithms: analytical calculation and mathematical simulation. In the first case, the characteristics of the algorithms are obtained in the form of functions of the model parameters and those of the measurement experiment. As a rule, the estimates of algorithm characteristics are received as a result of approximated analytical calculations, stipulating herewith the limitations on the parameters of models and an algorithm. In the second case, the estimates of the algorithms characteristics are obtained using a specific sets of values of model parameters and the measurement experiment, presented in the form of tables. The analytical calculation gives a more informative description of the metrological properties of the algorithm. However, in practice this approach is frequently rather difficult to realize because of the algorithm complexity. In this case, the mathematical simulation remains the only way out. In this investigation an analytical approach of evaluating the algorithm characteristics is realized.

4 Results of the Metrological Certification of Algorithms for Determination of the Location and Value of the Measurement Signal Extremum

Table 1 shows the estimates of the algorithm characteristics obtained for the indicated models of initial data errors (K(τ) -correlation function of noise, n(t)=δ_0 and n(t)=$\delta_0+\delta_1 t+\delta_2 t^2$, s - the area of the peak).) and for the model of the valid signal, characterized by the peak height and width, with the definite restrictions on the parameters of the initial data models and those of the filters (restrict.).

Table 2 presents the characteristics of algorithms for symmetric and asymmetric models of peaks, respectively. When estimating the algorithm characteristics a certain smoothness of the valid signal model (s(t)) is assumed, i.e. the existence of derivatives up to the fourth order and the variation from zero at the point of the second derivative extremum as well as the third one for asymmetric peaks and the fourth one for symmetric peaks.The following indices are used: h - the sampling interval, T - the interval during which the signal is processed (T=(N-1)h). In each cell of the table the first formula is absolute value of the estimation of location error charactristic, the second one - absolute value of the estimation value error characteristic. It should be mentioned that transformed error of the location of extrema is characterised by the limits, whereas expectation of transformed error of the value of extrema is always positive.

5 The Analysis of Metrological Certification Results

In conclusion, we would like to dwell on the analysis and the use of metrological certification results. In solving metrological problems, the final purpose lies in providing the required accuracy of the measurement result with the optimization of expenses for its realization, which is accomplished by planning the measurement procedure. The planning is fulfilled on the basis of a priori estimates of characteristics of the measurement result error with the definite relationships between the initial data parameters and those of the measurement experiment. The algorithm characteristics, obtained as a result of metrological certification, would be appropriate for use as the estimates a priori.

The analysis of the results of metrological certification of the algorithms shows the opposite character of the dependence of the methodical and transformed errors characteristics on the parameters of the algorithm and on those of the measurement experiment. It argues for existing of the rational values of the indicated parameters. The basic criterion for selecting the parameters is the provision of the required accuracy. At the same time, the additional conditions may be imposed, namely:

Table1

	RESTRICT.	K(τ)	LIMITS, CORRECTIONS	STANDARD DEVIATION
$h_1(\tau)$	$\frac{1}{\gamma} \ll T \ll D$	$\sigma^2 \exp\{-\gamma\|t\|\}$	$\Delta_h = \frac{T^2}{2} s''(t^*) + \delta_0$ $\Delta_t = T$	$\sigma_h = \frac{\sigma}{\sqrt{\gamma T}}$ $\sigma_t = \sqrt{\frac{4\sigma}{3\gamma T s''(t^*)}}$
$h_1(\tau)$	$\frac{1}{\gamma} \ll T \ll D$	$\sigma^2 \exp\{-\gamma^2 t^2\}$	$\Delta_h = \frac{T^2}{2} s''(t^*) + \delta_0$ $\Delta_t = T$	$\sigma_h = \sqrt[4]{\frac{\pi}{4}} \frac{\sigma}{\sqrt{\gamma T}}$ $\sigma_t = \sqrt{\sqrt[4]{\frac{\pi}{4}} \frac{4\sigma}{3\gamma T s''(t^*)}}$
$h_2(\tau)$	$\frac{1}{\gamma} \ll T \ll D$	$\sigma^2 \exp\{-\gamma\|t\|\}$	$\Delta_h = 2T^2 s''(t^*) + \delta_0$ $\Delta_t = 2T$	$\sigma_h = \frac{\sigma}{\sqrt{2\gamma T}}$ $\sigma_t = \sqrt{\frac{2\sigma}{3\gamma T s''(t^*)}}$
$h_2(\tau)$	$\frac{1}{\gamma} \ll T \ll D$	$\sigma^2 \exp\{-\gamma^2 t^2\}$	$\Delta_h = 2T^2 s''(t^*) + \delta_0$ $\Delta_t = 2T$	$\sigma_h = \sqrt[4]{\frac{\pi}{4}} \frac{\sigma}{\sqrt{2\gamma T}}$ $\sigma_t = \sqrt{\sqrt[4]{\frac{\pi}{4}} \frac{4\sigma}{3\gamma T s''(t^*)}}$
$h_3(t)$	$\frac{1}{\gamma} \ll T \ll D$ $\gamma \gg \alpha \gg \sqrt{\delta_2}$ $\gamma \gg \alpha \gg \delta_1$	$\sigma^2 \exp\{-\gamma\|t\|\}$	$\Delta_h = \frac{\alpha s}{2} + \frac{\delta_1 \alpha - 2\delta_2}{\alpha^2}$ $\Delta_t = T + \frac{\alpha h}{s''(t^*)}$	$\sigma_h = \frac{\sigma}{\sqrt{\gamma T}}$ $\sigma_t = \sqrt{\frac{4\sigma}{3\gamma T s''(t^*)}}$
$h_3(t)$	$\frac{1}{\gamma} \ll T \ll D$ $\gamma \gg \alpha \gg \sqrt{\delta_2}$ $\gamma \gg \alpha \gg \delta_1$	$\sigma^2 \exp\{-\gamma^2 t^2\}$	$\Delta_h = \frac{\alpha s}{2} + \frac{\delta_1 \alpha - 2\delta_2}{\alpha^2}$ $\Delta_t = T + \frac{\alpha h}{s''(t^*)}$	$\sigma_h = \sqrt[4]{\frac{\pi}{4}} \frac{\sigma}{\sqrt{2\gamma T}}$ $\sigma_t = \sqrt{\sqrt[4]{\frac{\pi}{4}} \frac{4\sigma}{3\gamma T s''(t^*)}}$

Table2

	LIMITS FOR SYMMETRIC PEAKS	LIMITS FOR ASYMMETRIC PEAKS	EXPECTATION	STANDARD DEVIATION
A	$0{,}016h^3\frac{x^{(4)}(t^*)}{x^{(2)}(t^*)}$, $0{,}0087h^4x^{(4)}(t^*)$	$0{,}04h^2\frac{x^{(3)}(t^*)}{x^{(2)}(t^*)}$, $0{,}17h^3x^{(3)}(t^*)$	$\frac{3h\sigma^2}{(x^{(2)}(t^*)h^2)^2}$, $\frac{\sigma^2}{x^{(2)}(t^*)h^2}$	$\frac{\sqrt{2}h\sigma}{x^{(2)}(t^*)h^2}$, σ
B	$0{,}22h^3\frac{x^{(4)}(t^*)}{x^{(2)}(t^*)}$, $0{,}21h^4x^{(4)}(t^*)$		$\frac{5{,}8h\sigma^2}{(x^{(2)}(t^*)h^2)^2}$, $\frac{2\sigma^2}{x^{(2)}(t^*)h^2}$	$\frac{\sqrt{2{,}5}h\sigma}{x^{(2)}(t^*)h^2}$, $\sqrt{0{,}51}\sigma$
C	$\frac{x^{(4)}(t^*)}{x^{(2)}(t^*)}(\bar{t}-t^*)*$ $*\left((\bar{t}-t^*)^2/3-0{,}016T^2\right)$, $x^{(4)}(t^*)\{0{,}13(\bar{t}-t^*)^4-$ $-0{,}021(\bar{t}-t^*)^2T^2-$ $-1{,}5*10^{-3}T^4\}$	$\frac{x^{(3)}(t^*)}{x^{(2)}(t^*)}*$ $*\left((\bar{t}-t^*)^2-\frac{T^2}{40}\right)$, $x^{(3)}(t^*)(\bar{t}-t^*)*$ $*\left((\bar{t}-t^*)^2-\frac{T^2}{40}\right)$	$\frac{720\sigma^2(\bar{t}-t^*)}{N(x^{(2)}(t^*)T^2)^2}$, $\frac{6\sigma^2}{Nx^{(2)}(t^*)T^2}*$ $\left(1+240\frac{(\bar{t}-t^*)^2}{T^2}\right)$	$\frac{7\sigma}{\sqrt{N}x^{(2)}(t^*)T^2}*$ $*\sqrt{T^2+60(\bar{t}-t^*)^2}$ $\frac{3\sigma}{\sqrt{N}}\sqrt{\frac{1}{4}-\frac{2(\bar{t}-t^*)^2}{T^2}+}$ $\overline{+\frac{20(\bar{t}-t^*)^4}{T^4}}$

insignificance of the methodical error as compared with the transformed one or insignificance of the random error as compared with the systematic one, and other conditions.

References

1. Yu.V.Tarbeyev, I.B.Chelpanov, T.N.Siraya, 2nd Symposium of the IMEKO Technical Committee on Metroloogy -TC8,-Abstr., 71 (1983).
2. I.B.Chelpanov, A.G.Ramazanova (Chunovkina),T.N.Siraya, Mera-90,-Abstr., v.III.
3. Muhammad A. Sharaf, Deborah L. Illman, Bruce R. Kowalski, Chemometrics, 1986, John Wiley and Sons, ISBN 5-7245-0361-1.

Advanced Mathematical Tools in Metrology III
Edited by P. Ciarlini, M.G. Cox, F. Pavese & D. Richter

A METHOD FOR EVALUATING TRENDS IN OZONE-CONCENTRATION DATA AND ITS APPLICATION TO DATA FROM THE UK RURAL OZONE MONITORING NETWORK

M. G. COX, P. M. HARRIS, M. J. T. MILTON and P. T. WOODS
National Physical Laboratory, Queen's Road, Teddington,
Middlesex, TW11 0LW, UK

The preliminary results of a study of long-term trends in the concentration of ozone measured at the 16 sites of the UK Rural Ozone Monitoring network are reported. The study used as its main tool a regressive model incorporating seasonal factors which is fitted to the ozone-concentration data. Particular statistics reported are the gradient with respect to time of the underlying straight-line trend in ozone concentration at each site, and a value for these gradient measures aggregated over all 16 sites in the network. The uncertainties of the statistics derived from the regressive model are estimated using bootstrap resampling methods. These methods have the advantage over traditional approaches that they do not require any assumption to be made about the sampling distribution of the data errors that are unexplained by the model.

1 Introduction

The growth in the extent and scope of air-quality monitoring has resulted in the accumulation of substantial archives of data measured at automatic stations. One of the most important questions that this data is intended to answer is whether ambient air quality is improving or deteriorating from year to year. Although this question appears to be straightforward, in practice long-term trends are difficult to quantify accurately because of the presence of large seasonal and random effects in the data.

In this paper, we specifically consider "raw data" consisting of hourly averages of ozone concentration measured at the 16 sites of the UK Rural Ozone Monitoring Network[4]. The sites include a variety of different rural locations such as remote hill tops, coastal areas and undeveloped areas between population centres. The ambient ozone concentration is recorded automatically and averaged over 15-minute intervals. It is then ratified, calibrated and "quality assured" prior to being placed on a public access database as one-hour averages. Although data is available at some sites from 1973, this study used a common six-year period between 1989 and 1994 to allow comparison of the results obtained at different sites and to permit aggregation of these results over the whole network. The questions addressed are (a) what is the long-term trend over the six-year period in ozone concentration at each site, and what is the statistical level of confidence in these estimates, and (b) how can the information from all 16 sites be aggregated to estimate the long-term trend over all sites, and what is the statistical level of confidence in this estimate?

Each of these questions is established as a formal statistical hypothesis and the results of our analysis are used to test them. The precise definition of "long-term change in ozone concentration" is crucial to the hypotheses and their testing. We use the arithmetic mean of the hourly data, computed over yearly, quarterly, monthly and weekly time periods, as measures of ozone concentration. An underlying straight-line trend in these measures, expressed as a *gradient* with respect to time, is used to indicate the long-term trend in ozone concentration.

2 Data fitting with regressive models

2.1 Straight-line regressive model

One approach to estimating the long-term trend at a site is to calculate annual averages of the data and then to fit a straight-line regressive model to these averages. Suppose y_i^*, $i = 1, \ldots, m$, denotes the value in parts-per-billion (ppb) of ozone concentration obtained by averaging (using the arithmetic mean) hourly data from year i. Then, the regressive model for these values takes the following form:

$$y_i^* = F_y(t_i, a, b_1) = at_i + b_1, \qquad t_i = i,$$

where a represents the gradient of the underlying straight-line trend in ozone concentration (measured in ppb per year), and b_1 is the value of ozone concentration at the start of the first year.

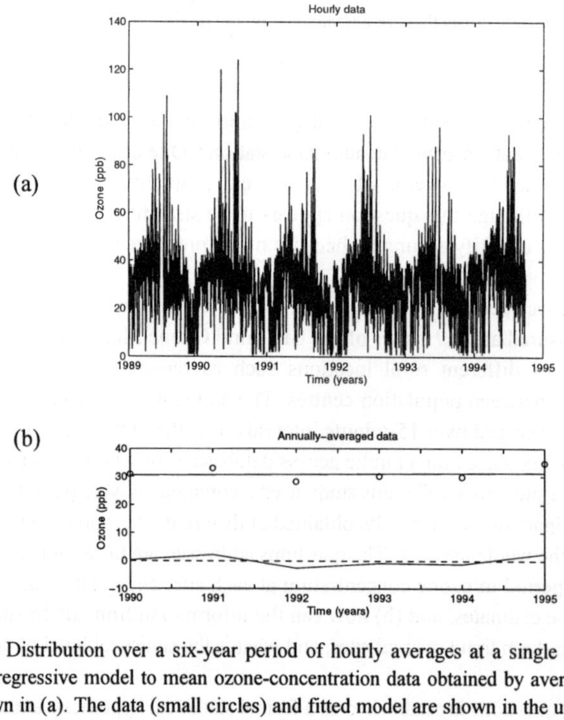

Figure 1: (a) Distribution over a six-year period of hourly averages at a single site. (b) Fit of straight-line regressive model to mean ozone-concentration data obtained by averaging annually the date shown in (a). The data (small circles) and fitted model are shown in the upper trace, and the weighted residuals (straight line joining their values) and "zero line" are shown below.

The parameters in the regressive model are determined by minimising the difference between the measured data values y_i and the model values y_i^* in a weighted least-squares sense, where the weights w_i are chosen to reflect the size of the sample that is averaged to yield y_i. An example of this approach for a single site is shown in Figure 1.

Although this approach has the advantage of simplicity, it has several disadvantages. Firstly, it "smooths out" much of the fine structure (and hence information) in the data. Secondly, it reduces a relatively large number of data points to a number for which only "small-sample" statistical theory is applicable. Finally, it does not provide realistic confidence intervals for the gradient (or any other modelled parameter) without making additional assumptions (such as that of normality). The residuals defined by the fitted model are not purely estimates of measurement error (which is two orders of magnitude smaller than a typical residual) but are predominantly quantities unexplained by the model which behave in a random or near-random manner.

2.2 Regressive models incorporating seasonal factors

Some of the drawbacks of the simple straight-line regressive model can be overcome by the use of a model that includes an underlying straight-line trend together with a set of factors, referred to as "seasonal factors", that account for periodic variations. There will be 1, 4, 12 or 52 factors according to whether yearly-, quarterly-, monthly- or weekly-averaged data is being analysed. They are not intended to explain the nature of the the periodic variations in the data, but are used as a statistical tool to remove some periodic trends in order to identify the underlying trend better.

Suppose $v_{i,j}^*, j = 1, \ldots, n$, are the values obtained by averaging over n equal time periods in year i. Then, the regressive model takes the form:

$$v_{i,j}^* = F_v(t_{i,j}, a, \boldsymbol{b}) = at_{i,j} + b_j, \qquad t_{i,j} = (i - 1) + \frac{j}{n}, \qquad j = 1, \ldots, n,$$

where a is the gradient of the underlying straight-line trend and $\boldsymbol{b} = (b_1, \ldots, b_n)^T$ are the seasonal factors. The parameters a and $\boldsymbol{b}$ in the regressive model are determined by minimising the difference between the data and model values in a weighted least-squares sense as described in Section 2.1.

3 Estimation of uncertainties

It is possible to use the regressive models described above to estimate standard uncertainties for the model parameters. A 95% confidence interval for a model parameter is given by $[x - ks, x + ks]$, where x is the estimated value of the parameter, s is its standard uncertainty, and k is a coverage factor[3]. The coverage factor k depends on the distribution of the weighted residuals: subject to the assumption that they are distributed normally with mean zero, k is approximately two and the confidence interval would be approximately $[x - 2s, x + 2s]$. However, in situations such as that discussed here, where it is not possible to make an assumption about the distribution of residual errors, we require an approach that is "distribution-free" or "non-parametric" in order to estimate confidence intervals. This is provided by *bootstrap resampling*[1,2].

Bootstrap resampling involves the repeated resampling of the residual deviations to generate new data sets that can be fitted with the regressive model to establish new estimates for the model parameters. In this application, if the parameters a defining the gradient of the underlying trend and $\boldsymbol{b}$ for the seasonal factors are computed as in Section 2 then we determine the model values $f_{i,j}$ and model weighted residuals $r_{i,j}$ from

$$f_{i,j} = F_{\nu}(t_{i,j}, a, \boldsymbol{b}), \quad \text{and} \quad r_{i,j} = w_{i,j}(\nu_{i,j} - f_{i,j}).$$

The values $r_{i,j}$ form a sample from the distribution of weighted residuals of the fitted model with respect to possible values $\nu_{i,j}$ for ozone concentration, and therefore provide a (discrete) approximation to this distribution.

A new (simulated) set of data $\nu_{i,j}^{+}$, generally different from the values $\nu_{i,j}$, is constructed by adding to each model value $f_{i,j}$ a random sample r taken from the set of values $r_{i,j}$(assuming replacement), *ie*, $\nu_{i,j}^{+} = f_{i,j} + r$. Here, r is termed a *bootstrap sample* from the set of residuals $r_{i,j}$. Having constructed the simulated data $\nu_{i,j}^{+}$, we may then process it in any way we choose. In particular, we may compute estimates a^{+} and $\boldsymbol{b}^{+}$ of the parameters in the model that fits this data in the manner described in Section 2.

This procedure is repeated a large number of times (N) to generate the $1 \times N$ matrix A of estimates a^{+} and the $n \times N$ matrix B whose columns are the estimates $\boldsymbol{b}^{+}$. Each *row* of these matrices contains a sample from the distribution for the corresponding parameter, and therefore provides a (discrete) approximation to this distribution. Since the elements in the $1 \times N$ matrix A form a sample from the distribution for the gradient in ozone concentration, the 2.5 and 97.5 percentiles of this empirical distribution specify a 95% confidence interval for the value of the trend.

In this way we may derive confidence intervals at any level for any of the model parameters without the need to make assumptions about the distribution of residuals associated with the model. In fact, we may obtain confidence intervals for any quantity *derived* from the model parameters. An example of this is described in Section 5 where methods for aggregating the results obtained from many sites are discussed.

4 Results for site data

The null hypothesis to be tested is that *there is no underlying straight-line trend over the time span of the data.* As it stands, this hypothesis cannot be tested without interpretation. Our interpretation is that it is identical to the hypothesis that *the gradient of the underlying long-term trend in the regressive model is zero.* For purposes of testing this hypothesis it is essential that the model is *validated.* This involves comparing with other models within the class, for example models with different numbers of seasonal factors (see below) or models including nonlinear trends or models applied to data averaged over time periods of different phase. The results of this validation, to be reported elsewhere, suggest that a model with a linear underlying trend applied to weekly-averaged data provides robust results.

The regresssive model described in Section 2.2 is fitted by weighted least-squares and the sampling distribution of the gradient of the underlying straight-line trend term empirically

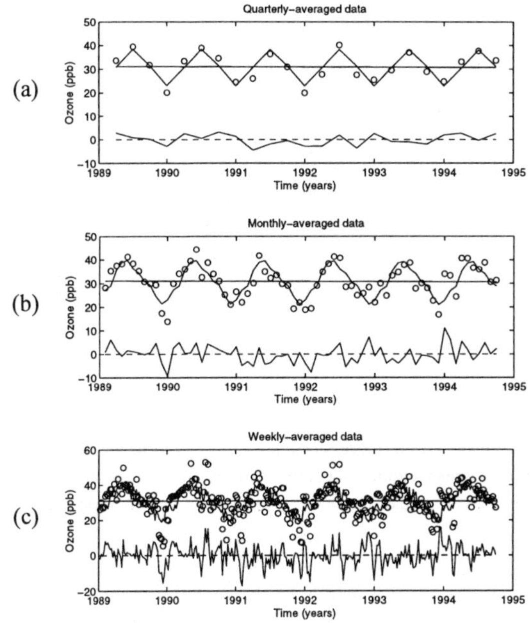

Figure 2: Fit of regressive model to mean ozone-concentration data obtained by averaging over (a) quarterly, (b) monthly and (c) weekly time periods for a single site. For each fit, the data (small circles), fitted model (straight line joining model values) and underlying trend are shown in the upper trace. The weighted residuals (straight line joining their values) and the "zero line" are shown below them.

determined using the non-parametric approach of bootstrap resampling. If the 95% confidence interval for the gradient, computed from this empirical distribution, does not contain zero, there is reason to doubt the hypothesis that ozone concentration is not changing. Figure 2 illustrates the fits to mean ozone-concentration data obtained by averaging over quarters, months and weeks the data shown in Figure 1(a). In Figure 3 we show graphically for all the sites and time periods, the estimates of the gradient together with their 95% confidence intervals obtained by bootstrap resampling. It is seen in this figure that, generally, the quarterly, monthly and weekly estimates are comparable, and the weekly results give the tightest confidence interval. "Annual" gradients are consistently higher than gradients for other periods. The results for the weekly-averaged data indicate that there is evidence to doubt the hypothesis for six of the 16 sites. Furthermore, for five of these six sites, the indication is that at the 95% confidence level, the trend in ozone concentration is increasing, whereas for the remaining site it is decreasing.

5 Estimating trends across all sites

The aim in this section is to consider how the results obtained for each site can be aggregated to

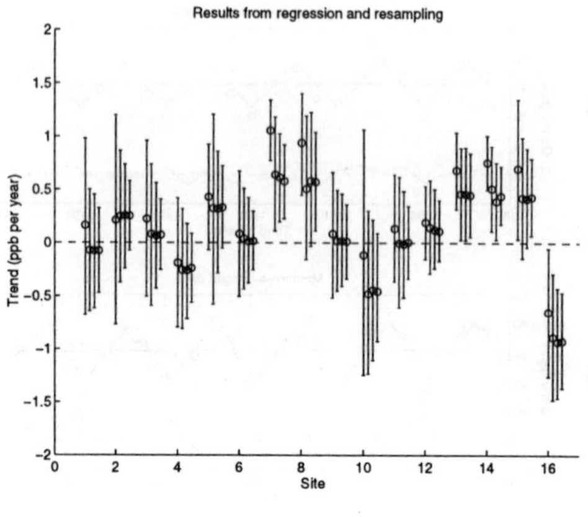

Figure 3: Best estimates (from regression) and 95% confidence intervals (from bootstrap resampling) for fits to mean ozone-concentration data for the 16 sites. For each site, results obtained by averaging over annual, quarterly, monthly and weekly time periods are shown from left to right.

evaluate the long-term trend over the whole network, and then to test the statistical hypothesis that *there is no long-term change in ozone concentration over the network.*

There is no single way of combining the results from the 16 sites to obtain a totally representative value for the whole network. Two possibilities include evaluating the mean of the individual site values, and counting the number of sites at which the long-term trend is increasing compared with the number at which it is not. Such statistics are straightforward to compute, but evaluating their corresponding significance is more difficult because only 16 values are available. We show here how bootstrap resampling can be used to overcome this problem.

As a result of the calculations described in Section 3, for each of the 16 sites, a matrix A_i, $i = 1, ..., 16$, of size $1 \times N$, that describe the sampling distribution of the gradient in ozone concentration was calculated. These matrices are arranged into a single $16 \times N$ matrix G that contains all N bootstrap estimates from all 16 sites. The array G of bootstrap estimates can form the basis for estimating the confidence interval associated with any estimator formed by taking a function (arithmetic mean, number of positive values, *etc*) of the gradients. For a specific estimator, we form its value for each column of the matrix G, *ie* for each bootstrap sample. The result is a $1 \times N$ matrix whose elements estimate the sampling distribution for that estimator. Then, the 2.5 and 97.5 percentiles of this empirical distribution specify a 95% confidence interval for the value of the estimator .

A straightforward approach to evaluating the trend over the whole network is to take the average of the trends at each of the sites. This is done by averaging the elements in the columns of the matrix G to form N bootstrap estimates of the network average. In Table 4 we list the average of the gradient values obtained from the original regressive fits to the data for each site (see Figure 3), that provides an estimate of the overall trend value, and its 95% confidence interval obtained from the frequency distribution. We note that the 95% confidence intervals are approximately symmetrically placed with respect to the least-squares estimates. For other statistics, such as the number of positive gradients, this is not the case. The results for weekly-averaged data, which we believe provide good information about the individual sites (Section 4), indicate that there is evidence to doubt the hypothesis that *there is no long-term change in the mean concentration of ozone over the whole network.*

Table 4: Estimates of, and 95% confidence intervals for, the mean long-term trend across all 16 sites.

Time period	Estimate	95% conf. int.
Annual	0.293	[0.133, 0.449]
Quarter	0.105	[-0.043, 0.250]
Month	0.094	[-0.026, 0.215]
Week	0.097	[0.010, 0.189]

6 Conclusions

Evaluating long-term trends in measured ozone concentration data is a difficult challenge for statistical analysis. We have proposed the use of regressive models with seasonal factors, together with bootstrap resampling from the distribution of model residuals obtained from the regression, to estimate the long-term trend in ozone concentration and its 95% confidence interval. The information provided by resampling can also be used to test hypotheses concerning data aggregated from many sites. The resampling technique gives a great deal more support to the estimated confidence interval since it does not assume that the measured data (or the mean trend at each site) are normally distributed. We believe the confidence interval calculated using resampling gives a good reflection of the underlying variability of the data.

References

1. Efron B. *The Jackknife, the Bootstrap, and Other Resampling Plans.* Society for Industrial and Applied Mathematics, Philadelphia (1982).
2. Efron B. and Tibshirani R. An Introduction to the Bootstrap. Monographs on Statistics and Applied Probablility 57. Chapman and Hall (1993).
3. ISO Guide to the Expression of Uncertainty in Measurement. The International Organization for Standardization (1993).
4. PORG. Ozone in the United Kingdom 1992. Third report of the United Kingdom Photochemical Oxidants Review Group, Harwell Laboratory, Oxfordshire, UK (1993).

Advanced Mathematical Tools in Metrology III
Edited by P. Ciarlini, M.G. Cox, F. Pavese & D. Richter

IDENTIFICATION OF SIGNAL COMPONENTS BY STOCHASTIC MODELLING IN MEASUREMENTS OF EVOKED MAGNETIC FIELDS FROM PERIPHERAL NERVES

T. GRÄBE[1], M. BURGHOFF[2], U. STEINHOFF[2], B.-M. MACKERT[3], G. CURIO[3]
[1]Technische Universität Berlin, Department of Electrical Engineering,
[2]Physikalisch-Technische Bundesanstalt, Berlin-Charlottenburg
[3]Universitätsklinikum Benjamin Franklin, Freie Universität Berlin, Department of Neurology

Using a low noise SQUID system, magnetic fields evoked by somatosensory activation of peripheral nerves were recorded over the lower spine. Averaging procedures have to be used to detect such weak signals. To perform weighted averaging, the variances and covariances of epochs were calculated. The distribution of the variances over the epochs allowed different noise components to be classified. The influence of disturbances in the recordings was reduced by weighted averaging.

1 Introduction

The evaluation of biological magnetic field measurements requires that different noise components be reduced. This reduction can be achieved by weighted averaging. The technique of weighted averaging has been found to be useful, for example, in the reconstruction of brainstem-auditory-evoked potentials [1]. We present and discuss additional possibilities of weighted averaging in the analysis of measurements of magnetic fields evoked by peripheral nerves.

2 Theory

2.1 Weighted averaging

Stimulation and magnetic field measurements are repeated N times. The description of the i-th epoch within each time interval after stimulation is

$$y_i(t) = s(t) + n_i(t), \quad i = 1,\dots,N, \tag{1}$$

where $s(t) := E[y_i(t)]$ is the evoked signal and $n_i(t)$ is the noise in terms of stationary random processes with variances

$$Var(n_i) = E\left[(n_i)^2\right] =: \sigma_i^2 \tag{2}$$

and covariances

$$Cov(n_i, n_k) = E[n_i n_k] =: \mu_{ik}\,. \tag{3}$$

No assumption is made about the spectral density or probability distribution. N is the number of epochs and $E[]$ is the expected value.

The weighted average $\hat{s}(t)$ of N measured epochs $y_i(t)$ is calculated as

$$\hat{s}(t) = \sum_{i=1}^{N} w_i y_i(t) \quad \text{with the normalization} \quad \sum_{i=1}^{N} w_i = 1. \tag{4}$$

Minimizing the mean square error $E\left[\left(s(t)-\hat{s}(t)\right)^2\right]$ considering correlated noise between epochs, the theoretical weights $w_{i,opt}$ and the corresponding minimal mean square error *mmse* are derived from equation (1):

$$\begin{pmatrix} \sigma_1^2 & \mu_{12} & \cdots & \mu_{1N} & -1 \\ \mu_{21} & & & & \\ \vdots & & \ddots & \vdots & \vdots \\ & & & & \\ \mu_{N1} & & \cdots & \sigma_N^2 & -1 \\ 1 & & \cdots & 1 & 0 \end{pmatrix} \begin{pmatrix} w_{1,opt} \\ \\ \vdots \\ \\ w_{N,opt} \\ mmse \end{pmatrix} = \begin{pmatrix} 0 \\ \\ \vdots \\ \\ 0 \\ 1 \end{pmatrix} \tag{5}$$

Note that some weights in (5) may be negative. In the special case of uncorrelated noise between epochs, (5) gives

$$w_{i,opt} = \frac{\frac{1}{\sigma_i^2}}{\sum_{k=1}^{N} \frac{1}{\sigma_k^2}} \quad \text{and} \quad mmse = \frac{1}{\sum_{k=1}^{N} \frac{1}{\sigma_k^2}}. \tag{6}$$

2.2 Variance estimation

To calculate weights according to (5) or (6) the variances and covariances are needed, which are unknown and must therefore be estimated from the measurements. This raises two problems:

1. The sample functions of the noise processes are not known, because measured epochs include evoked responses and noise processes.
2. Estimation of variances and covariances based on sample functions of a random process can be made in different ways. It depends on the properties of the measurements which way is appropriate.

Three strategies to bypass the first problem are discussed:

i. Epochs instead of noise sample functions are used to estimate the variances and covariances.

ii. The unweighted average is subtracted of before variances and covariances are estimated.

iii. ±-averaging is carried out (differences between two epochs).

2.2.1 Strategy i - Epochs instead of noise sample functions

Since the amplitudes of the signal s are very small compared with the noise process n, the unknown sample functions of the noise processes n_i are replaced by the measured epochs y_i to estimate the variances and covariances. As described in [1], the use of such variances in (6) can lead to an "underestimation" of the signal s in the weighted average. However, these variances or the corresponding weights can be used to analyse the noise processes and to classify epochs before averaging, for example, to detect large disturbances automatically. The use of such variances and covariances to calculate the weights according to equation (5) can dramatically intensify the effect of underestimating the signal s. This adulteration of the estimate is caused by the small signal s which is nevertheless present. The following strategies eliminate the influence of the signal s.

2.2.2 Strategy ii - Subtraction of the unweighted average

The variances and covariances are estimated by using instead of the measured epochs y_i, the modified epochs

$$\tilde{y}_i(t) = y_i(t) - \frac{1}{N}\sum_{k=1}^{N} y_k(t) \tag{7}$$

which do not contain the signal s but the mean value of all noise sample functions. When (5) is used, the weighted average comes close to the unweighted average. This effect cannot be observed when (6) is used.

2.2.3 Strategy iii - ±-averaging (differences between two epochs)

±-averaging is based on the relations

$$Var(n_1 - n_2) = Var(n_1) + Var(n_2) - 2Cov(n_1, n_2)$$

$$Var(n_1 + n_2) = Var(n_1) + Var(n_2) + 2Cov(n_1, n_2)$$

in the case of uncorrelated noise processes n_1, n_2 yielding

$$Var(n_1 - n_2) = Var(n_1 + n_2). \tag{8}$$

Since $n_1 - n_2 = y_1 - y_2$, equation (8) allows the noise variance estimate of the sum of two epochs to be computed on the basis of the difference between the same two epochs. For weighted averaging, the weight which is calculated from the difference between the epochs is assigned to the sum of the two epochs. The disadvantages of this method are

- bad variance estimates if the assumption about the correlation is not true
- there is no possibility of taking existing correlations into account and
- the theoretical quality of the average is deteriorated compared with the theoretical weighting of single epochs (half the measurements with twice the noise variances).

The use of ±-averaging is reasonable only if there are no correlations between the epochs.

2.2.4 Notes on the second problem, estimation of variances and covariances based on sample functions of a random process

The optimal variance estimation from L sample values $\underline{z} := [z(t_1), \cdots, z(t_L)]^T$ of an ergodic zero mean process $z(t)$ is

$$\hat{\sigma}^2 = \frac{1}{L} \underline{z}^T \underline{P}^{-1} \underline{z}, \tag{9}$$

where $\underline{P} = \frac{1}{\sigma_z^2} \underline{C}_{zz} = \frac{1}{\sigma_z^2} E\left[\underline{z}\,\underline{z}^T\right]$ is the normalized covariance matrix of the process $\{z(t_1), \cdots, z(t_L)\}$. The unknown theoretical variance σ_z^2 serves as normalization factor and is not contained in the normalized covariance matrix. $\underline{P}$ only carries the information about the form of the spectral density of the process, not about the value of the variance. Because the theoretical matrix $\underline{P}$ is unknown again, $\underline{P}$ would have to be estimated, leading to better variance estimates only if the number of sample values for the estimation of $\underline{P}$ is much larger than L, and premising stationariness of the spectrum over the whole time interval. An interpretation of (9) as a (time-varying) filter results from a decomposition (e. g. Cholesky decomposition) of $\underline{P}^{-1}$: $\hat{\sigma}^2 = \frac{1}{L} \underline{z}^T \underline{P}^{-1} \underline{z} = \frac{1}{L} \underline{z}^T \underline{F}^T \underline{F}\,\underline{z}$. In this case, $\underline{F}$ is called whitening filter, because the transformed process $\underline{F}\,\underline{z}$ has a white spectral density. In practice, the variance is estimated after digital filtering of the measured data. The filter is time-invariant and usually a highpass filter, since the noise processes naturally have lowpass spectra.

3 Application

3.1 Measurements

Using a multichannel SQUID system [3] inside a magnetically shielded room, we recorded 9000 epochs of evoked magnetic fields near the spine [2] using 49 magnetic field sensors covering a planar area of 210 mm diameter (analog bandpass: 2 Hz - 500 Hz, 2 kHz ADC). The length of one epoch was 55 ms. To evoke this nerve activity, we applied an electrical stimulus to the tibial nerve consisting of 9 Hz monophasic current pulses of 0,1 ms duration and 12 mA amplitude.

3.2 Results

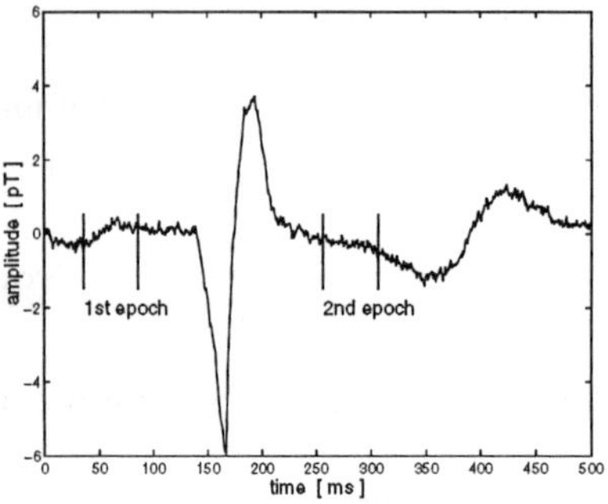

Figure 1: Example of one measurement signal in a time interval of 500 ms. The evoked signal of the nerve is completely covered by disturbances, e.g. the magnetic heart signal.

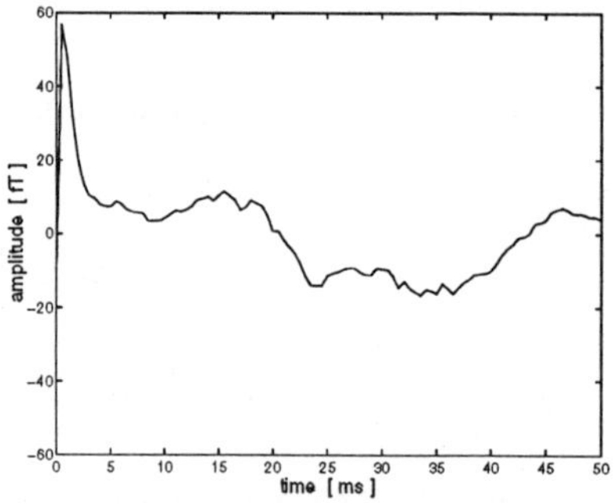

Figure 2: Unweighted average of 4500 epochs (same channel as figure 1). During the first 3 ms, the stimulus artefact is observable. The evoked nerve activity is expected between about 15 ms and 25 ms.

Figure 1 shows as an example one channel of the magnetic recording. Every epoch of the recorded data is composed of the evoked signal which is assumed to remain unchanged from epoch to epoch, and several noise components with up to 1000 times larger amplitude. The evoked nerve activity cannot be detected in the original recording without further noise reduction. The result of the method of noise reduction commonly used, unweighted averaging, is shown in Figure 2. The evoked nerve activity is expected from physiological predictions at about 15 ms to 25 ms after stimulation. Low-frequency disturbances in the average do not allow the evoked signal to be identified.

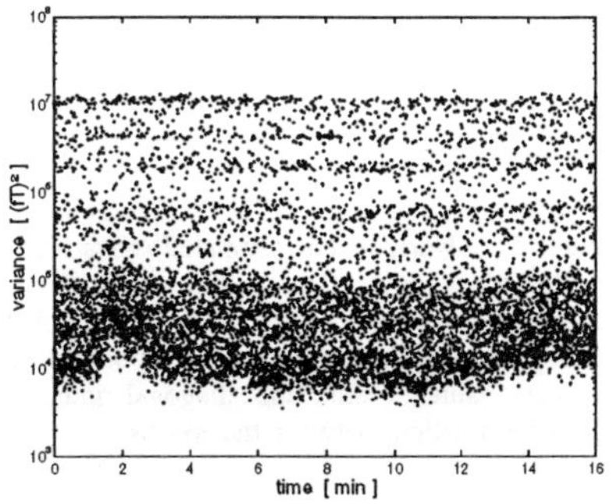

Figure 3: Estimated variances of the 4500 epochs without additional filtering show different noise groups.

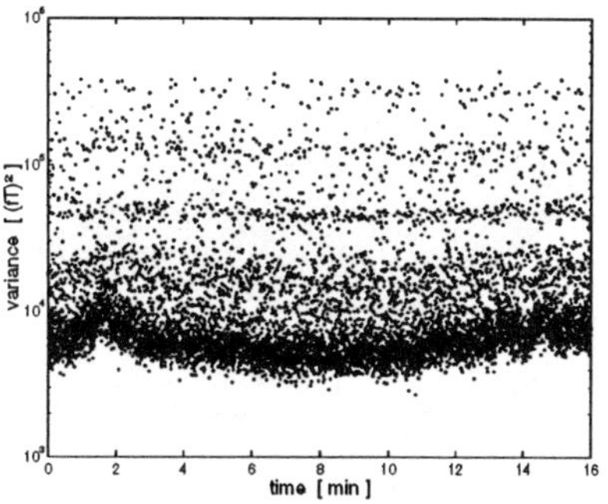

Figure 4: The use of a highpass filter before the estimation of the variances leads to a sharper clustering of some noise components.

Figure 3 shows the estimated variances for the above described strategy (i) using epochs instead of the noise. The variance estimates were computed according to equation (9) with $\underline{P} = \underline{I}$ (unity matrix). The variances clearly show different levels of noise, which can be clustered into different groups. The inspection of the recorded signal allows different noise level groups to be classified as belonging to different disturbances groups. The group with the lowest variance level belongs to epochs without strong heart signals. A slow variation of these variances corresponds to changing muscle tension of the volunteer during measurement. Other groups do not show this variation and correspond to different phases of the heart signal. Figure 4 shows that highpass filtering before variance estimation leads to a sharper variance group, corresponding to changing muscle tension. However, the influence of these different variance estimates on the resulting weighted average is small compared with the influence introduced when the covariances between epochs are taken into account.

Figure 5 shows the estimated covariances according to (5) for a block size of 50. Block size means the number of neighboured epochs whose correlations were considered. Higher values beside the diagonal occur only if significant correlation exists between the epochs. In our example clear correlations can be seen. Unless these correlations are taken into consideration during weighted averaging erroneous averages will result (see Figure 6). Consideration of these correlations during weighted averaging will prevent such erroneous results as can be seen in Figure 7. The low-frequency disturbances in the unweighted average can be reduced and the evoked signal detected. When this method is applied, the signal is

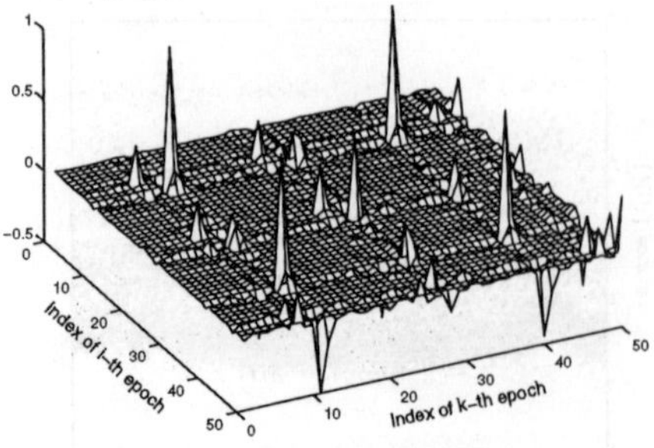

Figure 5: Normalized estimated covariance matrix for block size=50 according to (5). The values on the diagonal from (1,1) to (50,50) represent the normalized variances of epochs 1 to 50. Higher values beside the diagonal indicate significant correlations between the epochs.

underestimated. For a larger block size, e. g. a block size larger than 50, the weighted averaged signal of our example was underestimated to almost zero, whereas a block size smaller than 15 prevented underestimation.

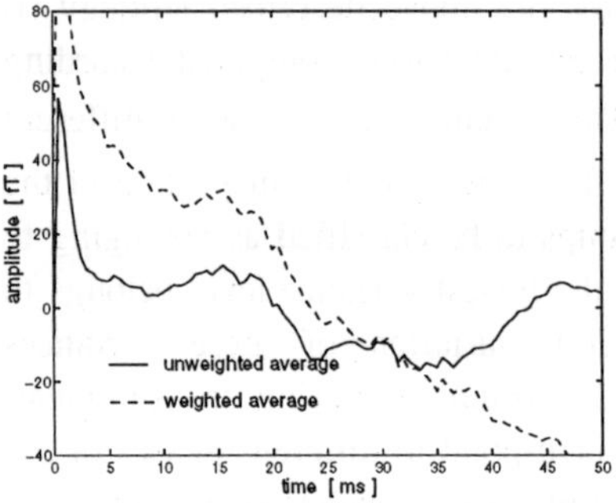

Figure 6: Weighted averaging without considering the existing correlation between the epochs leads to erroneous results.

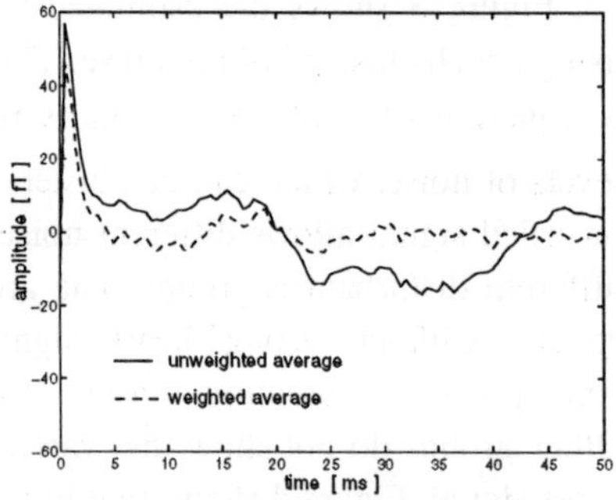

Figure 7: Weighted averaging with the correlation existing between the epochs taken into account leads to a slight underestimation of the signal and to clear discrimination of low-frequency noise components.

The use of the above strategy (ii), i. e. the variance and covariance estimates were computed from the epochs after subtraction of the unweighted average, shows that the weighted average approximates to the unweighted average itself and therefore brings about no improvement.

For the strategy (iii), i. e. ±-averaging, the estimation of the variances without consideration of the existing correlation has led to erroneous variance values. As a result, the average was almost as erroneous as the weighted average in which the correlations were left out of account (see Figure 6).

4 Discussion

The presentation of three strategies of weighted averaging procedures shows the sensitivity of weighted averaging. When weighted averaging is applied, the influence of disturbances on the average can be reduced, however, the effect of underestimation can be strong. For the block size of considered covariances between consecutive epochs a useful compromise has to be found between the underestimation and the erroneous results that are obtained if the correlations are neglected.

Beyond this, the examination of the weights of single epochs allows different noise components to be classified. Weighted averaging can reduce the influence of strong disturbances. The effect of some of these disturbances can be eliminated by conventional artefact rejection (thresholding), but this requires an adaptation to each signal and noise distribution within each channel of a multichannel system. The advantage of weighted averaging is the independence from a priori information about the characteristics of the disturbances. The relative weighting of the epochs against each other avoids the definition of threshold values for artefact detection.

If significant correlations between epochs exist in the recording, weighted averaging without consideration of these correlations leads to erroneous averages. ±-averaging, too, does not bring about an improvement when correlations exist.

We have presented an example of the usefulness of estimating variances for the classification of noise components in somatosensory evoked signals of peripheral nerves and of the usefulness of estimating covariances for weighted averaging.

5 References

[1] B. Lütkenhöner, M. Hoke, Ch. Pantev: Possibilities and Limitations of Weighted Averaging. Biol. Cybern. (1985) 409-416.

[2] G. Curio, B. Mackert, M. Burghoff, R. Kötitz, D. Drung, P. Marx: Neuromagnetic correlates of evoked somatosensory activity travelling through lumbosacral cauda equina and thoracic spinal cord in man. In: C. Baumgartner, L. Deecke, G. Stroink, S. J. Williamson (Editors), Biomagnetism: Fundamental Research and Clinical Applications Elsevier, 1995, 723-726.

[3] D. Drung: The PTB 83-SQUID-System for Biomagnetic Applications in a Clinic. IEEE Trans. Appl. Supercond. Vol. 5 (1995) 2112-2117.

Advanced Mathematical Tools in Metrology III
Edited by P. Ciarlini, M.G. Cox, F. Pavese & D. Richter

HIGH PRECISION 3D-CALIBRATION OF CYLINDRICAL STANDARDS

O. JUSKO, H. BOSSE, F. LÜDICKE

Physikalisch-Technische Bundesanstalt,
Laboratory "Dimensional Standards"
Bundesallee 100,
38116 Braunschweig, Germany

A procedure for calibration of cylindrical standards is described. As an example it is applied to a pressure standard of the piston-cylinder type. The procedure consists of measurement and numerical treatment of the object data. The measurements are performed by classical length- and form-measurement techniques. A numerical procedure generates three-dimensional carthesian co-ordinates of the cylinder surface from the measured raw-data. These co-ordinates are essential for simulation calculations of the pressure distribution in the pressure standard.

1 Geometrical data for the pressure simulation

The following procedure was developed for the calibration of pressure standards with respect to dimension and form, and will be described for this application. But it is not restricted to these special kind of standards and may be applied to all cylindrical objects of similar form quality.

Pressure standards of the piston-cylinder type are of major importance for high precision pressure gauging. They consist of a cylindrical piston, which can be slipped into a ring of nearly the same diameter, the pressure-cylinder. By application of suitable weights on top of the piston, the piston-cylinder-assembly nearly balances the pressure of a gaseous or liquid reservoir, which has to be calibrated. During measurement small movements of the piston are monitored, caused by small changes of the total volume and the pressure in the apparatus. These pressure changes may be predicted by finite-element method (FEM) based simulation calculations. For the application of FEM a precise geometrical description of piston- and cylinder-surface is necessary. From the known surface shapes the cross-section of the gap may be predicted. Such calculations are performed in the laboratory "Pressure" of the PTB [1]. The laboratory for "Dimensional Standards" is co-operating in generating the required co-ordinate data in carthesian form.

2 Form- vs. co-ordinate measurement techniques

Measurement of a cylinder surface is a standard application of a co-ordinate measurement machine (CMM). Hypothetically, all requested data could be measured in one setup of each component of the pressure standard. Therefore this solution path seems to be the most comfortable. But the drawback of CMMs is their comparatively high measurement uncertainty, which in best cases is of the order U = 1 μm (k=2) [2,3]. This value was validated by a recent internal comparison measurement [4]. For the FEM-based pressure calculation a much lower level of uncertainty is needed, about 0,050 μm. This uncertainty level can only be reached by separate specialized techniques of the classical dimensional metrology.

3 Geometrical measurements

Several pressure standards were measured recently. The one to be described here consists of a piston with a nominal diameter of 25 mm and about 60 mm length, and a pressure cylinder with ca. 40 mm active length and a nominal identical diameter.

To achieve the desired degree of measurement uncertainty, the geometrical measurements were performed with separate specialized instruments. These are diameter and form measurement instruments. The form measurements are here subdivided in roundness and straightness measurements. For each of those three measurement steps the calibration object had to be fixed on different mounting devices.

Principally form measurements only deliver relative co-ordinates, namely the radial or orthogonal relative displacement to a real or virtual metrological frame, which in most cases is realized by the guide of the measurement device. Without further information they cannot deliver a three-dimensional co-ordinate lattice.

3.1 Diameter measurements

At piston and cylinder, diameters were determined at two different measurement planes and in each plane in two orthogonal directions, symmetrically to a defined zero-point. For this purpose an one-dimensional Abbe-comparator, equipped with a laser-interferometer, was utilized. The achievable measurement uncertainty of this device amounts to U_D = 50 nm, where the main influence is caused by the calibration uncertainty of the diameter of the touching ball, which is used to mechanically contact the calibration object in the comparator [5,6]. The used touching forces are rather small, about 50 μN.

3.2 Straightness and parallelism measurements

For determination of the straightness deviations the z-axis of piston and cylinder of the pressure standard was aligned parallel to the x-axis of a universal measurement machine "Moore No. 3" and the data were measured in place.

Straighness deviations of 8 generatrices with an angular separation of 45° were determined, i.e. at $\phi = 0°, 45°, \ldots,$ and 315°. The straightness data were acquired with a spatial resolution of 0,01 mm with respect to the x-axis of the Moore instrument. To compensate the influence of the slideway deviations of the measurement device, the so-called reversal method was applied. The measurement uncertainty of the straightness deviations then amounts to $U_G =$ 0,03 μm. The measured deviations from straightness are in the range between 0,16 μm and 0,26 μm for the piston and between 0,19 μm and 0,52 μm for the cylinder (minimum zone straightness, gaussian-filter, limit wave-number 0,25 mm, 50 % transmission). The deviations from parallelism of opposite generatrices were also measured, but these are not essential for the further analysis. Their values are comparable to the straightness deviations.

3.3 Roundness measurements

The determination of the roundness deviations of the pressure standard were performed with a roundness measurement device "RTH Talyrond 73". The radial run-out of the spindle of this device, which amounts to approximately 0,030 μm, was compensated by application of the so-called multi-step method (10 orientations). The measurement uncertainty of roundness deviations of objects with high quality surface finish then amounts to $U_R = 0{,}010$ μm. If the surface finish is of lower quality, for example caused by an apparent waviness, as it is the case for most pressure standards and also for the standard under question, this value rises to 0,015 μm. [7 8 9]

The roundness deviations of piston and cylinder were determined at 11 different and equidistant planes, separated by a distance of 4 mm. The full circle was devided by 4096 equidistant measurement points. The measured deviations from roundness vary between 0,043 μm and 0,061 μm for the piston, and between 0,015 μm and 0,029 μm for the cylinder (minimum zone circle, 2RC-filter, limit wave number 50 U/R, 75 % transmission).

4 Data processing

As was emphasized in the last section, the form measurements are not sufficient for generating the three-dimensional co-ordinate lattice frame of the surface of the pressure standard. Additionally, diameter measurements have to be linked to the form measurements. For this purpose, an algorithm was developed and implemented with the programming language MODULA-2 running on a HP700 workstation under HP-UX. It will be described in the following.

4.1 Transformation of the measured raw-data to carthesian co-ordinates

It is assumed, that all data points of a roundness measurement belong to a single plane perpendicular to the z-axis of the pressure standard. This assumption is valid after thorough parallel alignment of the z-axis of the calibration object to the c-axis of the roundness measurement device.

The straightness deviations of the generatrices are assumed to be describable only by x,y-components, and that the z-coordinate is given by the index of the measurement point i.e. also in this case nearly perfect alignment is assumed.

The procedure starts at the roundness measuremnt planes with diameter measurements. The roundness trace of this plane is centered with radius $R = 0$. From this values and the two diameter values in the directions $0° - 180°$ and $90° - 270°$ within this plane corrected diameters are calculated. These are averaged to a mean "cylinder nominal radius", which is used to transform the roundness values. This adaptation of the diameter values is motivated by the different measurement uncertainties of diameter and roundness measurements. This procedure is applied for the second diameter plane in the same way.

The new radius values at the points of intersection of the diameters with the generatrices are further used as reference points, which help to tilt the generatrices to a corrected orientation. At last, the roundness measurements of all measurement planes are least-squares fitted to the reference points.

Thus, the carthesian co-ordinates are generated as follows: The z-coordinate of the roundness data-points is given by the position of the measurement plane at the cylinder-surface. The relative co-ordinate, corresponding to the original transducer deflection, and the corrected cylinder nominal radius (in μm) are totaled, resulting in a value r_0.

The x-co-ordinate of a data point of a roundness measurement is given by:

$$r_0 \cdot cos(\frac{2 \cdot \pi \cdot (n-1)}{4096})$$

Accordingly, the y-co-ordinate is given by:

$$r_0 \cdot sin(\frac{2 \cdot \pi \cdot (n-1)}{4096})$$

The index of a measurement point corresponds to a certain measurement height at the calibration object.

Table 1: x-, y-components of the generatrices after transformation to three-dimensional carthesian co-ordinates

position ϕ	x	y
0°	r_0	0
90°	0	r_0
180°	$-r_0$	0
270°	0	$-r_0$
45°	$r_0 \cdot \sqrt{2}/2$	$r_0 \cdot \sqrt{2}/2$
135°	$- r_0 \cdot \sqrt{2}/2$	$r_0 \cdot \sqrt{2}/2$
225°	$-r_0 \cdot \sqrt{2}/2$	$-r_0 \cdot \sqrt{2}/2$
315°	$r_0 \cdot \sqrt{2}/2$	$-r_0 \cdot \sqrt{2}/2$

The z-co-ordinates of the generatrices are sequentially given by their index. Again, an alignment of the pressure standard to the measurement axis, which in this case is the x-axis of the measurement device, is assumed. An influence of the orientation correction is neglected, because the corresponding angle is very small. The transducer deflections of the generatrices measurements and the corrected cylinder nominal radius (in μm) are totaled, resulting in individual values r_0. Table 1 lists the resulting x,y-components of the co-ordinates for selected data points.

4.2 Fit-contradictions

The reference points were measured with the roundness and also with the straightness measuring instrument i.e. there is principally redundant information. Possible deviations of both values are fit-contradictions. It can serve as a criterion for the quality of the measurements, for example with respect to the alignment on the different devices, and as a test for the synthesis-algorithm. The contradictions should be smaller than the measurement uncertainty. In our case they amount to 0,035 μm maximum with standard deviations below 0,015 μm.

4.3 Parameters of the fitted cylinder

Additionally to the individual co-ordinates a global parameter is calculated, the deviation from cylindricity deviation. For this purpose the "general method" of Murthy et.al. is applied [11] . The cylindricity is a useful value for the check of the production quality of the pressure standard. In our case it amounts to

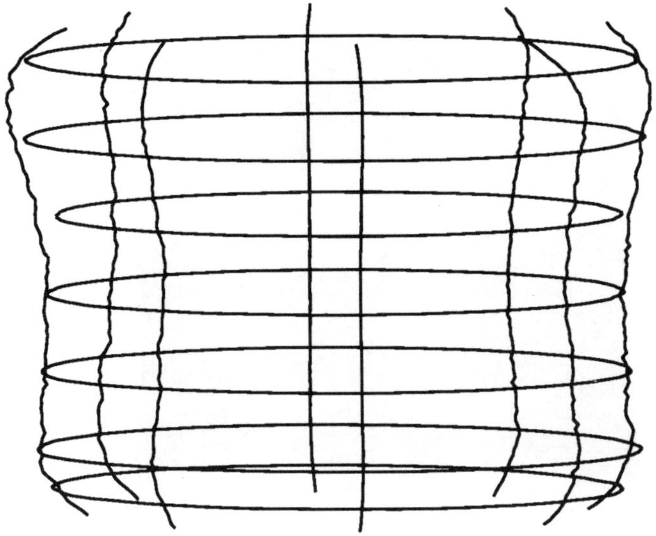

Figure 1: Visualization of cylinder-data, "bird-cage"-view

0,21 μm for the piston with a standard deviation of the fitted points from the calculated mean cylinder of 0,03 μm. The corresponding values for the cylinder of the pressure standard are 0,31 μm and 0,06 μm. The calculated cylinder radii of the fitted cylinders are $R_1 = 12491,908\ \mu m$ and $R_2 = 12492,563\ \mu m$, where the greater value corresponds to the pressure cylinder. Thus, the diffence of both cylinder radii amounts to only 0,655 μm. Yet, the both parts can be paired.

4.4 Visualization

The generated data can be made visible by the help of special 3D-visualization software. For this purpose a small routine has been developed in the programming language IDL[10] running on HP-UX. It allows to display the data-points, given in 3D carthesian coordinates, either in a "bird-cage" plot as shown in fig.1 or as gouraud-shaded surface-plot as shown in fig.2.

In fig.1 a view is selected, that makes fit-contradictions visible. They can

Figure 2: Visualization of cylinder-data, shaded view

be seen as not-touching of generatrices and roundness traces, especially on the left side of the plot. The plot is not drawn to scale, but, for the purpose of visualization, the real radius was discarded, the deviations from perfect shape were rescaled to a visible value, and an arbitrary plot-radius was used.

For the display of fig.2 principally the same procedure was applied. But as there is real data only on the roundness and generatrix traces, the missing areas were filled by interpolation. Although this procedure seems to add non-existent information, it sometimes helps to better understand the three-dimensional shape of a given calibration object.

5 Results and 3D-data measurement uncertainties

Deriving an error budget is not an easy task. The comments of this section may be seen as a first attempt. Yet, the given values are assumed to be calculated conservatively i.e. as a maximum estimation.

5.1 *Uncertainty of the position of a generatrix with respect to the cylinder-axis*

Contributions to the position uncertainty of a generatrix do result from the uncertainties of the diameter differences in both reference planes and the uncertainties of the straightness data-points nearby those planes. The uncertainty of the touching-ball diameter does not contribute to the generatrix position. Taking this into account, an uncertainty of $U_P = 0,035\ \mu m$ for the generatrices positions results.

5.2 *Uncertainty of the radial distance of a measurement point from the cylinder-axis*

Contributions to the radial distance uncertainty come from the uncertainty of the overall position of the fitted co-ordinate frame, the uncertainty of the absolute distance to the cylinder-axis, and the uncertainties of the original relative co-ordinates (transducer deflections) of the roundness and straightness measurements. The uncertainty of a single profile measurement point can be estimated to 0,015 μm maximum. Taking this value and the position uncertainties of the generatrices into account, an overall uncertainty of 0,045 μm for the 3D-cylinder radii co-ordinates results.

5.3 *Comparison with pressure measurements*

The generated data were taken as input data for a FEM pressure simulation. Within the limits of the uncertainty of the pressure measurement a perfect agreement of the simulation data to the measurement was found [12]. This corresponds to a relative uncertainty of $4 \cdot 10^{-6}$ with respect to the area of clearance of a selected measurement plane (z-co-ordinate) at the standard.

6 Conclusions and outlook

The achieved measurement uncertainty remains under the initial aim of 0,050 μm and the quality of the generated data was validated by independent pressure simulations and measurements. Thus, it can be stated, that the presented procedure is able to deliver three-dimensional co-ordinate data with uncertainties, which are about 15 to 20 times lower than possible with CMMs. Drawbacks are a very laborious and time-consuming multi-step measurement procedure and a similar complex evaluation. Therefore, in the current state, it will only be applicable to special very demanding calibrations.

A current new development of a multi-purpose measurement machine might help to shorten at least the measurement steps and may even lower the overall measurement uncertainty[13].

References

1. W. Sabuga, *Elastic distortion calculations at PTB on LNE 200 MPa : pressure balances as part of EUROMET Project 256*, PTB-Report W-63, 1995
2. Measurement uncertainties are calculated according to: *Guide to the Expression of Uncertainty in Measurement*, ISO 1995
3. H. N. Hansen, and L. De Chiffre, *Nordic Audit of Coordinate Measuring Machines*, Final Report of EUROMET Project No. 330, 1996
4. E. Trapet, private communication
5. F. Lüdicke, and H.-J. Rademacher, *Bestimmung von Maß und Form an Zylindern*, PTB-Mitteilungen, p. 99 (1989) 6, p. 429-433
6. M. Schaller, and H.-J. Rademacher, *Calibration of Standards for Industrial Metrology*, Acta Imeko V (1982) , p. 481
7. K. Thiele, *Normalmeßverfahren für Kreisformverkörperungen*, Metrologische Abhandlungen, 8 (1993) 3, p. 225-229
8. D. J. Whitehouse, *Some theoretical aspects of error separation techniques in surface metrology* , J. Phys. E.: Sci. Instr, 9 (1976), p. 531-536
9. C. Linxiang, *The measurement accuracy of the multistep method in the error separation technique* , J. Phys. E.: Sci. Instr. 22 (1989), p. 903-906
10. *Interactive Data Language* , trademark of Research Systems Inc.
11. T. S. R. Murthy, and S. Y. Rao, *A Simple Approach for Evaluation of Cylindrical Surfaces*, Annals of the CIRP 30 (1981), p. 441-444
12. J. Jäger, and W. Sabuga, private communication
13. M. Neugebauer, F. Lüdicke, and H. Bosse, *Ein neuer Komparator zur Maß- und Formkalibrierung an Zylindern, Kugeln und Quadern unter Reinraumbedingungen*, to be published

Advanced Mathematical Tools in Metrology III
Edited by P. Ciarlini, M.G. Cox, F. Pavese & D. Richter

MAGNETIC DIPOLE ESTIMATIONS FOR MCG-DATA

EKKEHARD KRAUSE [1]

Department of Mathematics, University of Greifswald, Germany

The analysis of heart diseases by the electrocardiogram (ECG) has been a standard method for decades. In the past 10 years the development of SQUIDs has made a reliable analysis using measurements of the magnetic field of the human heart (MCG) possible. During certain periods of the heart cycle MCGs look like dipoles. Therefore MCGs may be approximated by the equivalent current dipole and by the equivalent magnetic dipole which is defined by five and six parameters, respectively. The dipole parameters are usually determined using a method based on least squares fit.
We establish a new algorithm which calculates the magnetic dipole parameters from the measured data in almost real time and therefore in the future may give rise to fast procedures for both data reconstruction and analysis.

1. Introduction

In the analysis of heart diseases the electrocardiogram (ECG) has been one of the major tools for decades. In addition to that in the future the magnetocardiogram (MCG) may be useful in examining patients with heart diseases. The MCG does not describe the body surface potential as the ECG but the z-component of the magnetic vector field $\mathcal{B}(x, y, z)$ generated by the human heart during its excitation.
The measurement device of the PTB-laboratory in Berlin-Charlottenburg uses 37 SQUIDs. As illustrated in figure 1, when the data are recorded, the patient is lying on his back and the SQUIDs are localised in a frontal plane

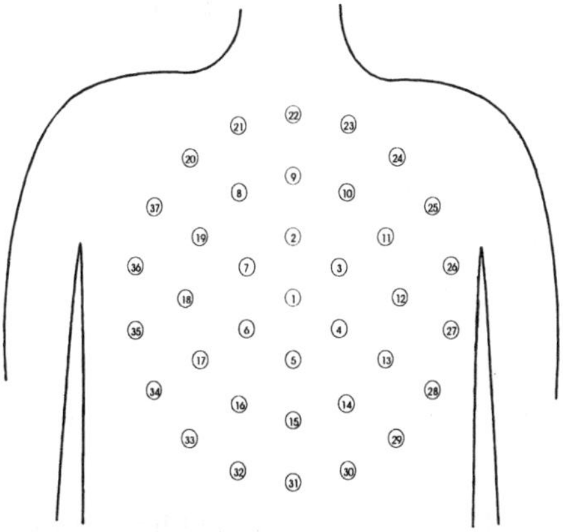

Figure 1: Geometry of the measurement device at the PTB-laboratory in Berlin-Charlottenburg

[1] I would like to thank M.Burghoff and U.Steinhoff of the MCG-group of the PTB in Berlin for providing the data and helpful discussions as well as my supervisor C.Bandt, M.Borchardt and S.Wilhelmi of Greifswald University for supporting this work.

directly over the chest of the patient. This plane is the $(x, y, z = 0)-$plane of our coordinate system, where the positive $x-$axis points to the left hand side of the patient and the positive $y-$axis to the head of the patient. The positive $z-$axis points into the patient's frontal direction.

A magnetic dipole is the first term of the magnetic multipole expansion. This means that the magnetic field of any current distribution looks like a magnetic dipole field when looked at from a long distance compared to the extension of the current distribution. Therefore, if there is only a small area of the human heart excited, a magnetic dipole may be a good approximation to the magnetic field of the human heart. On the other hand, if this is not the case, the magnetic dipole may be still very useful in addition to the equivalent current dipole in order to solve the inverse problem [1].

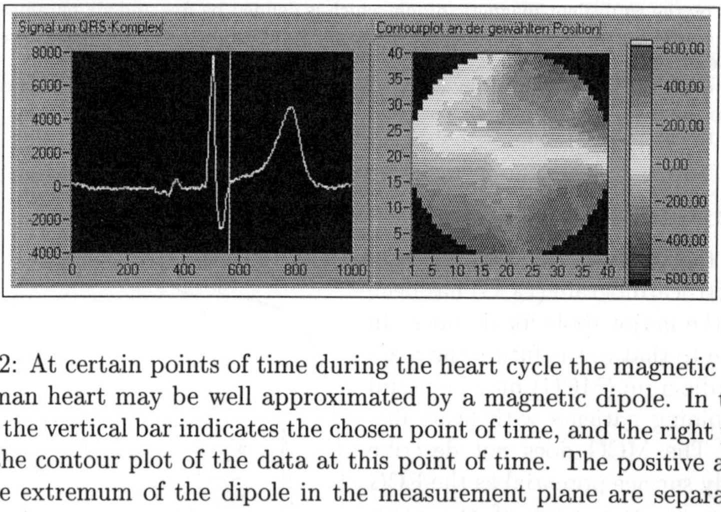

Figure 2: At certain points of time during the heart cycle the magnetic field of the human heart may be well approximated by a magnetic dipole. In the left picture the vertical bar indicates the chosen point of time, and the right picture shows the contour plot of the data at this point of time. The positive and the negative extremum of the dipole in the measurement plane are separated by the (light) zero line. During the heart cycle both extrema move in a certain way, so in order to analyse MCGs it seems to be promising to investigate this movement.

The formula obtained for a magnetic dipole at the point $\vec{r_0}$ with the magnetic dipole moment $\vec{m}$ is equivalent to the formula describing the electric dipole field [2], that is

$$\mathcal{B}(\vec{r_0}, \vec{m}; \vec{r}) = \frac{\mu_0}{4\pi} \cdot \frac{3\Big(\vec{m} \cdot (\vec{r} - \vec{r_0})\Big)(\vec{r} - \vec{r_0}) - \|\vec{r} - \vec{r_0}\|^2 \vec{m}}{\|\vec{r} - \vec{r_0}\|^5}, \tag{1}$$

where $\vec{r_0} = (x_0, y_0, z_0)$, $\vec{m} = (m_x, m_y, m_z)$ and $\vec{r} = (x, y, z)$.

In the next sections we shall introduce a new technique for calculating the six parameters of a magnetic dipole. We will consider the magnetic dipole as lying in a homogeneous conducting half–space. Our method is not based on least squares fit [1] and allows time analysis of MCG–signals using the magnetic dipole model in almost real time. Please note that a similar but easier procedure may be employed in order to obtain the five parameters of the equivalent current dipole (the sixth parameter of the equivalent current dipole is not determined by the $z-$component of the magnetic vector field).

2. The Magnetic Dipole Model

As mentioned above, MCG–data have a dipole structure during certain periods of the heart cycle. The recorded data describe the $z-$component of the magnetic field of the human heart measured in the $(x,y)-$plane.

The field of a magnetic dipole which is *not* contained in the $(x,y)-$plane has a positive and a negative extremum in this plane, unless the dipole moment is perpendicular to the $(x,y)-$plane. Figure 3 shows the field distribution of a magnetic dipole in the plane $(x,y,0)$ which is located at $\vec{r_0} = (0,0,-6)$ and has magnetic moment $\vec{m} = (\cos\frac{\pi}{8}, 0, \sin\frac{\pi}{8})$. The idea is to approximate the magnetic field of the human heart as shown in figure 2 by the field of one magnetic dipole as shown in figure 3. From formula (1) we see that a magnetic

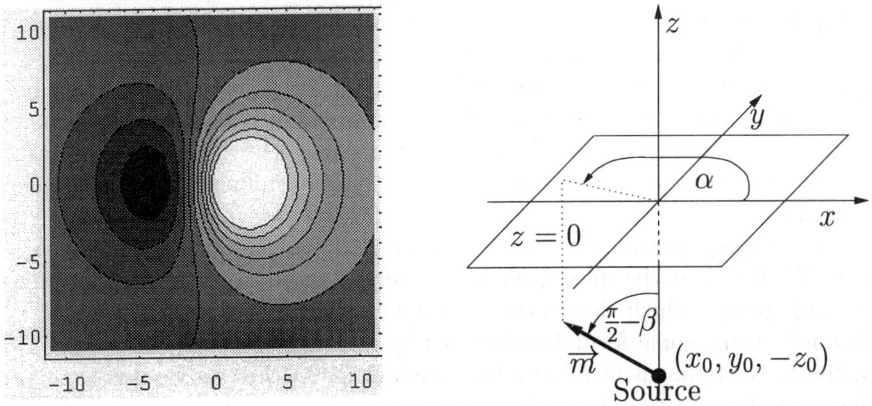

Figure 3: **Left:** Magnetic dipole at $\vec{r_0} = (0,0,-6)$ with magnetic moment $\vec{m} = (\cos\frac{\pi}{8}, 0, \sin\frac{\pi}{8})$, that is $\beta = \frac{\pi}{8}$ and $\alpha = 0$. **Right:** The SQUIDs of the measurement device are arranged in the plane $(x,y,z=0)$.

dipole is defined by the point $\vec{r_0} \in \mathbb{R}^3$ at which it is located and by its magnetic dipole moment $\vec{m} = (m_x, m_y, m_z)$. We parametrise $\vec{m}$ by the angles $\alpha \in [0, 2\pi]$ and $\beta \in [-\frac{\pi}{2}, \frac{\pi}{2}]$ and by the norm of $\vec{m}$, where α denotes the angle of $\vec{m}$ in the $(x, y)-$plane with the positive $x-$axis and β the angle of inclination of $\vec{m}$ towards the $(x, y)-$plane, that is

$$m_x = C\cos\alpha\cos\beta \tag{2}$$
$$m_y = C\cos\beta\sin\alpha \tag{3}$$
$$m_z = C\sin\beta, \tag{4}$$

where the constant C is defined by

$$C = \frac{\mu_0}{4\pi}\cdot\sqrt{m_x^2 + m_y^2 + m_z^2}. \tag{5}$$

The next formula describes the $z-$component of the magnetic dipole field defined by equation (1), which we shall use in the following:

$$B_z[(x_0, y_0, -z_0), (\alpha, \beta, C), (x, y, 0)] =$$
$$\frac{3\,C\,z_0\big((x - x_0)\cos\alpha\,\cos\beta + (y - y_0)\cos\beta\,\sin\alpha + z_0\sin\beta\big)}{\|\vec{r} - \vec{r_0}\|^5} - \frac{C\sin\beta}{\|\vec{r} - \vec{r_0}\|^3} \tag{6}$$

3. The Calculation of the Magnetic Dipole Parameters

In this section we show how the six parameters of the magnetic dipole can be calculated without using a method based on least squares fit.

To establish our formulas we need to assume that the magnetic field is known in the whole $(x, y, 0)-$plane. However, the measurement area covers only a bounded area. One way to overcome this problem is to establish formulas which use different powers of the measured data. By this procedure values significantly different from zero are amplified and the influence of smaller values is neglected.

The dipole parameters $\vec{r_0}$, α, β and C are estimated in the following way: We find first the maximum and minimum positions of $B_z(x, y, 0)$ in the $(x, y, 0)-$plane, which determine the angle α and the y_0-component of $\vec{r_0}$. Secondly, using some formulas obtained by integration we calculate the angle β, the z_0-component of $\vec{r_0}$ and the constant C. At the end we use β, z_0 and the positions of both extrema to calculate x_0.

In order to calculate z_0 and β independently on the positions of the extrema and to obtain C we integrate powers of B_z over the $(x, y)-$plane and define

$$I_j[z_0, \beta, C] = \int\limits_{\varphi=0}^{2\pi}\int\limits_{r=0}^{\infty}\big(B_z[(x_0, y_0, -z_0), (\alpha, \beta, C), (r\cos\varphi, r\sin\varphi, 0)]\big)^j\, r\, dr d\varphi\,. \tag{7}$$

The numerical values of $\{I_j[c, \beta, C] \,|\, j = 4, 5, 6\}$ are evaluated from the measured data. The following formulas show the solutions of integral (7) for $j = 4$, 5 and 6. We solve the system of nonlinear equations

$$I_4[z_0, \beta, C] = \frac{9\,\mathrm{C}^4\,\pi}{z_0^{10}} \frac{(713 - 644\cos 2\beta + 51\cos 4\beta)}{8960} \tag{8}$$

$$I_5[z_0, \beta, C] = \frac{3\,\mathrm{C}^5\,\pi}{z_0^{13}} \frac{(778418\sin\beta - 216621\sin 3\beta + 13601\sin 5\beta)}{1352078} \tag{9}$$

$$I_6[z_0, \beta, C] = \frac{3\,\mathrm{C}^6\,\pi}{z_0^{16}} \Big(\frac{1040510 - 1232651\cos 2\beta}{2050048} + \frac{243698\cos 4\beta - 12677\cos 6\beta}{2050048}\Big) \tag{10}$$

and obtain first for β the equation

$$\begin{aligned} K\,&\left(-9.104\,{I_5}^2 + 8.714\,I_4\,I_6\right) \\ &+ K^2\,\left(10.52\,{I_5}^2 - 10.07\,I_4\,I_6\right) + K^3\,\left(-5.721\,{I_5}^2 + 5.484\,I_4\,I_6\right) \\ &+ K^4\,\left(1.456\,{I_5}^2 - 1.395\,I_4\,I_6\right) + K^5\,\left(-0.139\,{I_5}^2 + 0.133\,I_4\,I_6\right) \\ &= -2.995\,{I_5}^2 + 2.865\,I_4\,I_6\,, \end{aligned} \tag{11}$$

where $K = \cos^2\beta$. In order to obtain equation (11) we had to perform non–unique calculations. For this reason equation (11) gives only the modulus of β but not its sign. Looking at equation (9) we see, that I_5 and β have equal signs. To get the right solution for β we therefore define

$$\beta[I_4, I_5, I_6] = \begin{cases} \arccos(\sqrt{K}) & \text{if } I_5 > 0 \\ -\arccos(\sqrt{K}) & \text{if } I_5 \le 0\,. \end{cases} \tag{12}$$

For simplification we introduce

$$\delta = 713 - 644\cos 2\beta + 51\cos 4\beta \tag{13}$$

$$\epsilon = 778418\sin\beta - 216621\sin 3\beta + 13601\sin 5\beta \tag{14}$$

and obtain as solutions for the constant C and z_0 for $\beta \neq 0$

$$C[I_4, I_5, \beta] \approx 1.652 \cdot 10^{-9} \cdot \left(\frac{I_4}{\delta}\right)^{\frac{13}{2}} \cdot \pi^{-\frac{3}{2}} \cdot \left(\frac{\epsilon}{I_5}\right)^5 \tag{15}$$

$$z_0[I_4, I_5, \beta] \approx 1.539 \cdot 10^{-4} \cdot \left(\frac{I_4}{\delta}\right)^{\frac{5}{2}} \cdot \pi^{-\frac{1}{2}} \cdot \left(\frac{\epsilon}{I_5}\right)^2 \tag{16}$$

In the case $\beta = 0$ the integral I_5 in (9) is equal to zero. We therefore use equations (8) and (10) and obtain C and z_0 dependent on I_4 and I_6 :

$$C[I_4, I_6] \approx 0.657 \cdot I_4^4 \cdot I_6^{-\frac{5}{2}} \quad (17)$$

$$z_0[I_4, I_0] \approx 0.767 \cdot I_4^{\frac{3}{2}} \cdot I_6^{-1}. \quad (18)$$

In order to determine the x_0- and y_0-component of the dipole location we think of a transformed coordinate system where the $x-$axis goes through both extrema. With respect to this coordinate system $y_0 = 0$ and x_0 is somewhere in between the extrema. To get the exact value of x_0 we think of a dipole located at $(0, 0, -z_0)$, calculate the positions of its extrema and assign them to the positions of the extrema obtained from the data. The positions of both extrema are roots of the derivative of $B_z[(0, 0, -z_0), (0, \beta, C), (x, 0, 0)]$ as defined in (6) with respect to x

$$\frac{3\,C\,\left(\sin\beta\, x^3 - 4\, z_0\, \cos\beta\, x^2 - 4\, z_0^2\, \sin\beta\, x + z_0^3\, \cos\beta\right)}{(x^2 + z_0^2)^{\frac{7}{2}}}. \quad (19)$$

For $\beta \neq 0$ we obtain as roots of (19)

$$x_1[z_0, \beta] = \frac{4z_0}{3}\left(\cot\beta + \cos\left(\frac{A(\beta)}{3}\right)\,(3 + 4\cot^2\beta)^{\frac{1}{2}}\right) \quad (20)$$

$$x_2[z_0, \beta] = \frac{4z_0}{3}\left(\cot\beta + \cos\left(\frac{A(\beta)}{3} + \frac{2\pi}{3}\right)\,(3 + 4\cot^2\beta)^{\frac{1}{2}}\right) \quad (21)$$

$$x_3[z_0, \beta] = \frac{4z_0}{3}\left(\cot\beta + \cos\left(\frac{A(\beta)}{3} + \frac{4\pi}{3}\right)\,(3 + 4\cot^2\beta)^{\frac{1}{2}}\right), \quad (22)$$

where

$$A(\beta) = \arccos\left(\frac{\cot\beta\,\left(117 + 128\cot^2\beta\right)}{16\,(3 + 4\cot^2\beta)^{\frac{3}{2}}}\right).$$

In the case $\beta = 0$ the two zeros of (19) are given by $\pm\frac{z_0}{2}$. For $\beta \neq 0$ it can be shown that $x_3[z_0, \beta]$ is the position of the extremum with the bigger modulus and $x_1[z_0, \beta]$ or $x_2[z_0, \beta]$ is the position of the second extremum if $\beta < 0$ or $\beta > 0$, respectively.

4. Reconstruction using Sampled Data

In the previous section we have introduced formulas for calculating the six parameters of a magnetic dipole. These formulas are based on two assumptions: Firstly, we need to know the z−component of the magnetic dipole field in one plane. Secondly, we should know B_z at every point in this plane. Unfortunately neither the first nor the second assumption is satisfied in practice. The reason for this is that we measure the data only in a bounded area, which leads to a truncation error, and on the other hand we measure the field only at certain sample points, that is there is a discretisation error.

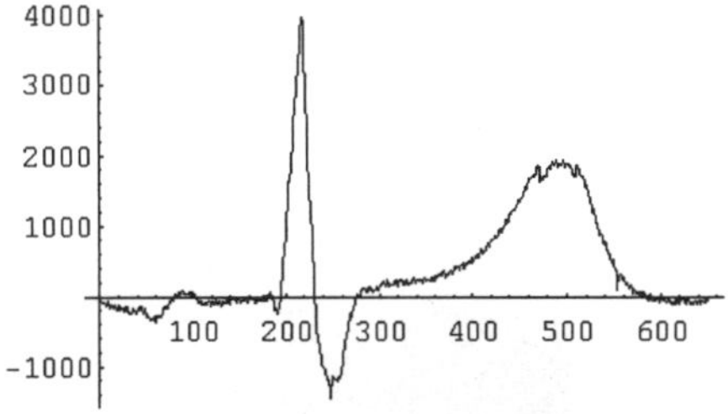

Figure 4: This MCG–signal of a healthy patient was recorded at position 1 (cf. figure 1) and shows one heart cycle.

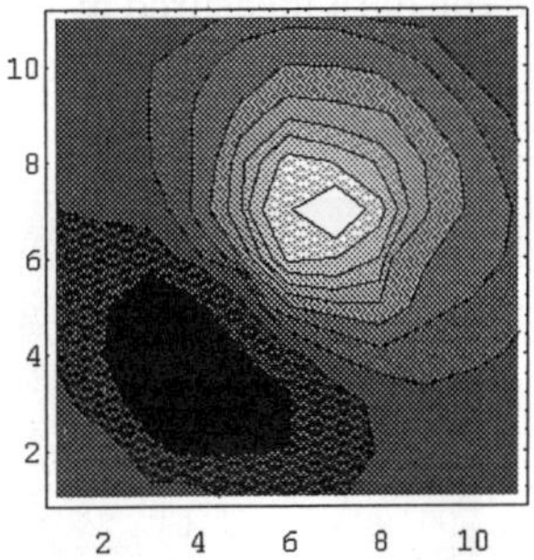

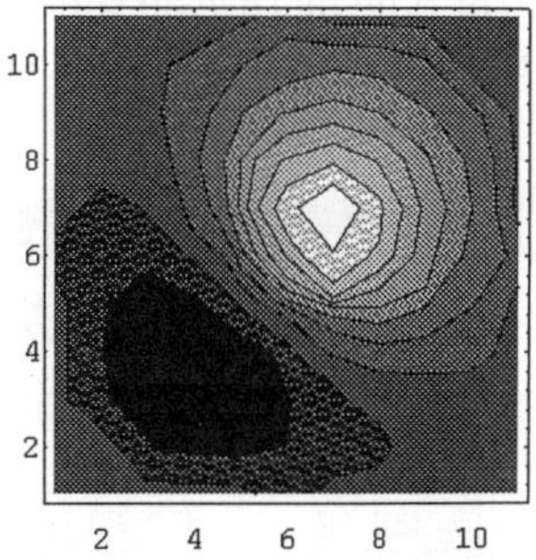

Figure 5: The picture on the left side shows the field distribution of the magnetic dipole $\vec{r_0} = (0, 0, -7)$, $\alpha = \frac{2\pi}{7}$, $\beta = 0.5$ and the constant $C = 15 \cdot 10^6$ in the plane $(x, y, 0)$. The result of the reconstruction can be seen in the picture on the right side.

Clearly, the more sample points we use the better the reconstruction gets. Therefore we restrict ourselves to the measurement design shown in figure 1 and check, whether the method works for as few as 37 sample points. For this reason we calculate the magnetic dipole field at the sample points shown in figure 1 and calculate the reconstruction using the obtained dipole parameters.

As mentioned in the introduction MCG–data have a dipole structure only during certain periods of the heart cycle such as parts of the QRS–complex. Figure 4 shows the MCG–signal of a healthy patient measured at position 1. The contour plot of this MCG resembles a dipole field for instance during the periods of time from 210 to 244 and from 450 to 600. Figures 6 and 7 show the contour plot of the measured data and of the reconstructed data at the end of the R–peak and in the middle of the T–wave, respectively. Although we have used at this stage only the equivalent magnetic dipole and not in addition to that the equivalent current dipole as suggested in [1], the fit of the data is not too bad.

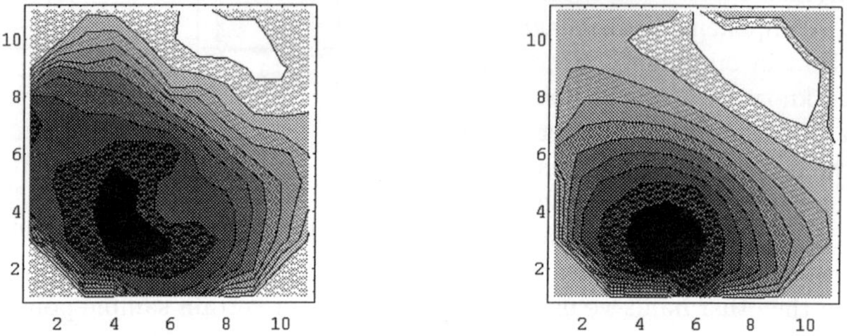

Figure 6: The picture on the left–hand side shows the data measured at position 244 (R–peak), the picture on the right–hand side shows the reconstructed data.

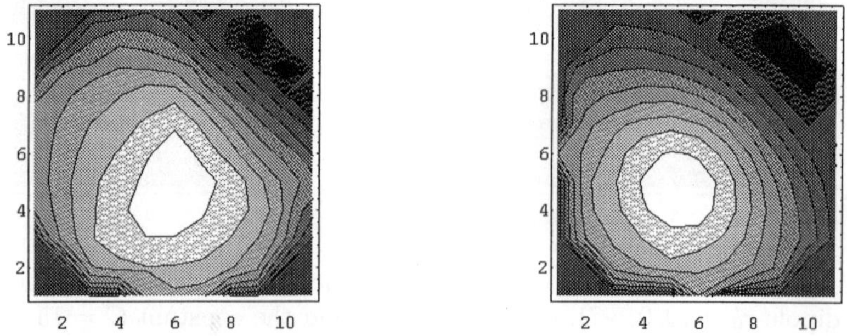

Figure 7: The picture on the left–hand side shows the data measured at position 505 (T–wave), the picture on the right–hand side shows the reconstructed data.

References

[1] R.S. GONELLI and M. AGNELLO. Inverse problem solution in cardiomagnetism using a current multipole expansion of primary sources. *Phys. Med. Biol.*, Vol. 32, No. 1, 133-142, 1987.

[2] W. GREINER. *Theoretische Physik Band 3 – Klassische Elektrodynamik.* Verlag Harri Deutsch, 1991.

Advanced Mathematical Tools in Metrology III
Edited by P. Ciarlini, M.G. Cox, F. Pavese & D. Richter

TRANSFER FUNCTIONS OF DISCRETE SPLINE FILTERS

M. KRYSTEK

Physikalisch-Technische Bundesanstalt, Bundesallee 100,
D-38116 Braunschweig, Germany

Abstract

Due to the imperfection of the measurement process, filtering is mandatory to remove noise before measurement results can be evaluated. This kind of filtering is also known as smoothing of the data set. Such smoothing can be done by curve fitting, if an analytical description of the data is known and applicable. But often this is not possible. In this case it is appropriate to use splines as a substitute. If the data are given as a discrete set, discrete splines have to be used. This smoothing process can then be considered as the application of a discrete (digital) low-pass filter. Such filters are usually characterized by transfer functions. It will therefore be shown how the transfer functions of discrete spline filters can be constructed.

1 Introduction

Due to the imperfection of the measurement process measured data points are usually given only approximately. Therefore it is mandatory to remove any noise by a certain kind of filtering, also known as smoothing of the data set, before measurement results can be evaluated. If the form of the underlying function describing the data is known a priori, it is recommended to approximate the latter by an appropriate trial function which is fitted to the data points by application of the well-known least squares technique. But often such a function is not known. In this case, it is appropriate to use splines as a substitute. If the data are given as a discrete set, discrete splines have to be used. Such a smoothing process can then be considered as the application of a discrete (digital) low-pass filter having a certain transfer function and impulse response function. From the point of view of signal analysis it is therefore interesting to know, how transfer functions and impulse response functions of discrete spline filters can be constructed.

Although the method shown in this paper is applicable to any kind of discrete splines, only the simple case of discrete periodic polynomial splines will be dealt with here in order to keep the mathematics simple.

2 Discrete periodic polynomial splines

The measured data $y(x)$ as a function of the variable x are usually given by the measurement process at n equidistant discrete points $x_k = k\,\Delta x$, $(k = 0 \ldots n-1)$. Filtering of these data normally should allow a certain degree of smoothness to be achieved. Owing to this assumption, Whittaker [1] suggested as early as 1923 that statistically satisfactory results can be obtained by appropriately balancing a tradeoff between infidelity to data and infidelity to a smoothness constraint imposed upon $s(x_k)$, $(k = 0 \ldots n-1)$, being the resulting data when the filtering process has been applied. The infidelity to data is usually expressed by a least mean squares criterion as

$$\sum_{k=0}^{n-1} [y(x_k) - s(x_k)]^2 \,, \tag{1}$$

while the smoothness is guaranteed when a member of the family of spline functions is fitted through the data. Since the data are discrete, discrete splines (cf. [2] and [3]) are likely to be a good choice. The simplest member of this spline family is the discrete polynomial spline of degree m, obtained by minimizing the expression

$$\sum_{k=0}^{n-1} [\nabla^m s(x_k)]^2 \,, \tag{2}$$

with integer m and the operator ∇ of the backward differences defined by

$$\nabla s(x_k) = s(x_k) - s(x_{k-1}). \tag{3}$$

Thus a suitable filter can be constructed by minimizing the functional expression

$$F\left(s_0, s_1, \ldots, s_{n-1}; p, m\right) = \sum_{k'=0}^{n-1} \left(\nabla^m s_{k'}\right)^2 + p \sum_{k'=0}^{n-1} \left(y_{k'} - s_{k'}\right)^2 , \tag{4}$$

with respect to the s_k, $(k = 0 \ldots n-1)$, where the abbreviations $s_k = s(x_k)$ and $y_k = y(x_k)$ have been used. The parameter p controls the tradeoff between infidelity to data and infidelity to smoothness. In addition s_k has to fulfill the periodic boundary conditions

$$s_{k+n} = s_k, \qquad k = 0 \ldots n-1, \tag{5}$$

because $y(x)$ is periodic by definition, so $s(x)$ should be periodic, too.

As usual, the minimum of $F\left(s_0, s_1, \ldots, s_{n-1}; p, m\right)$ is obtained from the conditions

$$\frac{\partial}{\partial s_k} F\left(s_0, s_1, \ldots, s_{n-1}; p, m\right) = 0, \qquad k = 0 \ldots n-1, \tag{6}$$

yielding

$$\sum_{k'=0}^{n-1} \nabla^m s_{k'} \frac{\partial}{\partial s_k} \nabla^m s_{k'} - p(y_k - s_k) = 0, \qquad k = 0 \dots n-1. \tag{7}$$

By solving the recurrence relation $\nabla^m s_{k'} = \nabla(\nabla^{m-1} s_{k'})$

$$\nabla^m s_{k'} = \sum_{i=0}^{m} (-1)^i \binom{m}{i} s_{k'-i}, \qquad k' = 0 \dots n-1, \tag{8}$$

can be obtained, and thus

$$\sum_{k'=0}^{n-1} \nabla^m s_{k'} \frac{\partial}{\partial s_k} \nabla^m s_{k'} = \sum_{i=0}^{m} \sum_{i'=0}^{m} (-1)^{i+i'} \binom{m}{i} \binom{m}{i'} s_{k+i'-i} = \tag{9}$$

$$\sum_{l=-m}^{m} (-1)^l \binom{2m}{m+l} s_{k+l}, \qquad k = 0 \dots n-1,$$

is valid. The latter identity has been obtained by rearranging the terms of the double sum. Inserting relation (9) into equation (7) finally yields

$$s_k + \frac{1}{p} \sum_{l=-m}^{m} (-1)^l \binom{2m}{m+l} s_{k+l} = y_k, \quad k = 0 \dots n-1. \tag{10}$$

The difference equation (10) together with the periodic boundary conditions (5) define a discrete periodic polynomial spline of degree m. Moreover, this is also the defining equation of a recursive filter transforming the input data set y_k into the output data set s_k.

3 Transfer functions

Filters are unequivocally determined by their transfer functions. To get the transfer function of the discrete polynomial spline filters, the discrete Fourier transform will be used.

The discrete Fourier transform $\tilde{Y}_j$ of a data set y_k is defined by the following pair of transformation equations:

$$\tilde{Y}_j = \sum_{k=0}^{n-1} y_k \exp\left(\mathrm{i}\frac{2\pi}{n} j\, k\right) \tag{11}$$

$$y_k = \frac{1}{n} \sum_{j=0}^{n-1} \tilde{Y}_j \exp\left(-\mathrm{i}\frac{2\pi}{n} j\, k\right). \tag{12}$$

where $i = \sqrt{-1}$, as usual. The angular frequency ω is given by

$$\omega = \frac{2\pi}{n\Delta x} j = j, \tag{13}$$

with Δx being the sampling interval. Thus $\tilde{Y}_j$ may also be expressed as $\tilde{Y}(\omega)$.

Applying the discrete Fourier transform to the filter equation (10) yields after some manipulations

$$\tilde{S}(\omega) = \tilde{H}(\omega)\tilde{Y}(\omega), \tag{14}$$

with $\tilde{Y}(\omega) = \tilde{Y}_j$ and $\tilde{S}(\omega) = \tilde{S}_j$ being the discrete Fourier transforms of y_k and s_k, respectively, and

$$\tilde{H}(\omega) = \left[1 + \frac{1}{p} \sum_{k=-m}^{m} (-1)^k \binom{2m}{m+k} \exp\left(-\mathrm{i}k\,\omega\Delta x\right)\right]^{-1} \tag{15}$$

being the transfer function of the discrete polynomial spline filter of degree m. By use of the identity (28) given in the appendix, equation (15) can be simplified to

$$\tilde{H}(\omega) = \left[1 + \frac{1}{p}\left(2\sin\frac{\omega\Delta x}{2}\right)^{2m}\right]^{-1}. \tag{16}$$

As can be seen from equation (16), $\tilde{H}(\omega)$ is always real. Thus, discrete polynomial spline filters are phase-correct filters, i.e. the filtering process does not alter the phase of the signal. Moreover, replacing j by $j \pm n$ in equation (13) adds $\pm 2\pi$ to the argument $\omega\Delta x$. This does not change the value of the sine in equation (16), and so does not change $\tilde{H}(\omega)$ either. Hence, $\tilde{H}(\omega)$ is periodic with the same period as y_k and s_k. Further, it can be deduced that the discrete spline filters are low-pass filters, because $\tilde{H}(0) = 1$ follows from equation (16).

To come closer to the traditional form of the transfer function, the parameter p will now be replaced by the cut-off value ω_c and the parameter α defined by

$$\tilde{H}(\omega_c) = \frac{1}{1+\alpha^2}. \tag{17}$$

From equations (16) and (17)

$$p = \frac{1}{\alpha^2}\left(2\sin\frac{\omega_c\Delta x}{2}\right)^{2m} \tag{18}$$

and

$$\tilde{H}(\omega) = \left[1 + \alpha^2\left(\frac{\sin\dfrac{\omega\Delta x}{2}}{\sin\dfrac{\omega_c\Delta x}{2}}\right)^{2m}\right]^{-1} \tag{19}$$

can then be obtained. The transfer function $\tilde{H}(\omega)$ according to equation (19) is shown in figure 3. As can be seen from this figure, the transfer function becomes steeper with increasing parameter m.

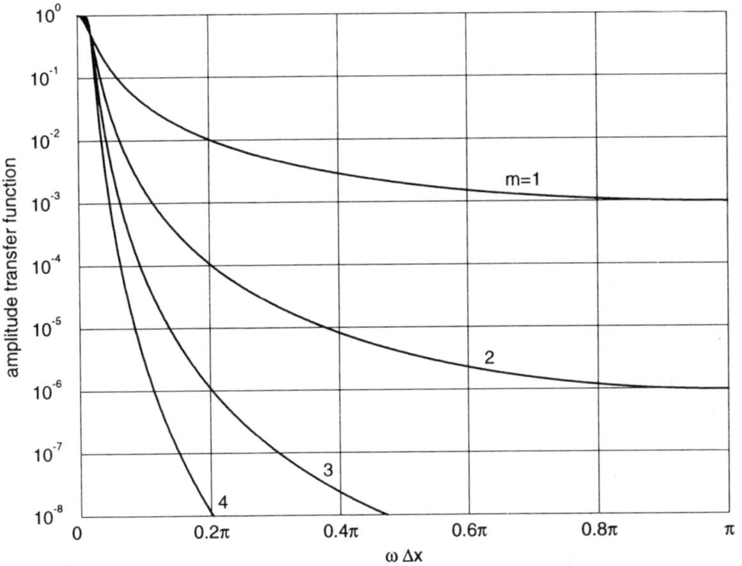

Fig. 1: Amplitude transfer function $\tilde{H}(\omega)$ of the discrete periodic spline filters for different parameters m ($\alpha = 1$, $\omega_c \Delta x = \pi/50$).

When the number of data points n is supposed to increase to infinity, i.e Δx approaches zero, $y(x)$ becomes continuous, and from equation (19) it follows that the transfer function $\tilde{H}(\omega)$ changes to

$$H(\omega) = \lim_{\Delta x \to 0} \tilde{H}(\omega) = \left[1 + \alpha^2 \left(\frac{\omega}{\omega_c}\right)^{2m}\right]^{-1} . \qquad (20)$$

This is the square of the absolute value of the transfer function of the well known Butterworth filter of order m as used in signal processing.

4 Impulse response function

In addition to their transfer function filtets are usually characterized by their impulse response function $\tilde{h}(x) = h_k$, ($x = k\Delta x$). This function can be obtained from the inverse Fourier transform of the transfer function

$\tilde{H}(\omega)$ as follows. Using equations (16), (12) and (13) and the identity $e^{iu} = \cos u - i \sin u$,

$$h_k = \frac{1}{n}\sum_{j=0}^{n-1}\left[1+\frac{1}{p}\left(2\sin\frac{\pi}{n}j\right)^{2m}\right]^{-1}\cos\frac{2\pi}{n}kj \tag{21}$$

$$-\frac{i}{n}\sum_{j=0}^{n-1}\left[1+\frac{1}{p}\left(2\sin\frac{\pi}{n}j\right)^{2m}\right]^{-1}\sin\frac{2\pi}{n}kj$$

is obtained. The second sum in equation (21) can easily be shown to vanish. Thus one finally gets

$$\tilde{h}(x) = \frac{1}{n}\sum_{j=0}^{n-1}\left[1+\frac{1}{p}\left(2\sin\frac{\Delta x}{2}j\right)^{2m}\right]^{-1}\cos xj. \tag{22}$$

As would be expected, the impulse response function $\tilde{h}(x)$ is real, even and periodic.

5 Example

Smoothing of measured data by filters is usual, as e.g. in roundness measurement. The filter prevailing today, recommended for use for this purpose is the so called phase corect 2RC filter. As shown in [4], this type of filter can be approximated by a discrete periodic spline filter with $m = 1$ and $\alpha = 1/\sqrt{3}$. Thus the resulting transfer function according to equation (19) is

$$\tilde{H}(\omega) = \left[1+\frac{1}{3}\left(\frac{\sin\dfrac{\omega\Delta x}{2}}{\sin\dfrac{\omega_c\Delta x}{2}}\right)^{2}\right]^{-1} \tag{23}$$

and the impulse response function yielded according to equations (18) and (22) is

$$\tilde{h}(x) = \frac{1}{n}\sum_{j=0}^{n-1}\left[1+\frac{1}{3}\left(\frac{\sin\dfrac{\Delta x}{2}j}{\sin\dfrac{\Delta x}{2}\omega_c}\right)^{2}\right]^{-1}\cos xj. \tag{24}$$

It can be shown, that the impulse response function can also be written as

$$h_k = \tanh\frac{b}{2}\exp(-b|k|), \qquad k = 0,\dots,n-1 \tag{25}$$

with

$$b = 2\log\left(\sqrt{3}\sin\frac{\omega_c\Delta x}{2} + \sqrt{1 + 3\sin^2\frac{\omega_c\Delta x}{2}}\right) \tag{26}$$

If Δx is sufficiently small, h_k and b, respectively, may be expanded into Taylor's series, yielding

$$h_k = \frac{\sqrt{3}}{2}\omega_c\Delta x\exp(-\sqrt{3}\,\omega_c\Delta x|k|) + O(\Delta x^3), \qquad k = 0,\ldots,n-1. \tag{27}$$

Equation (27) gives the discrete representation of the impulse response function of the phase correct 2RC filter. Thus the discrete periodic spline with $m = 1$ and $\alpha = 1/\sqrt{3}$ can be used instead of this filter type.

A C program implementing the spline filter according to equation (10) and (18) in the preceding sections has been applied to a measured profile. The result is shown in fig. 2. The computing time on a SUN 10/40 was 100 ms for 4096 data points. This is an order of magnitude faster than the usual phase-correct 2RC filter algorithm, even still if compared with the FFT version of the 2RC filter.

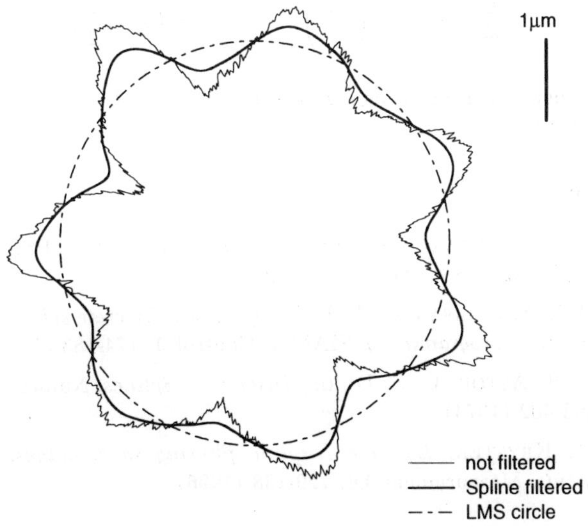

Fig. 2: Measured roundness profile of a ground rod filtered by a discrete periodic spline filter. Parameters: $n = 4096$, $m = 1$, $\alpha = 1/\sqrt{3}$, $\omega_c = 8$ undulations per revolution.

Appendix

In this appendix it will be shown, that

$$\sum_{k=-m}^{m} (-1)^k \binom{2m}{m+k} e^{-ikx} = \left(2\sin\frac{x}{2}\right)^{2m} \tag{28}$$

is valid. Using the well known identity $e^{i\alpha} - e^{-i\alpha} = i\,2\sin\alpha$ one gets

$$(1-e^{-ix})^{2m} = \left[e^{-i\frac{x}{2}}(e^{i\frac{x}{2}} - e^{-i\frac{x}{2}})\right]^{2m} = \left[e^{-i\frac{x}{2}}\, i\,2\sin\frac{x}{2}\right]^{2m} = \tag{29}$$

$$i^{2m} e^{-imx}\left(2\sin\frac{x}{2}\right)^{2m} = (-1)^m e^{-imx}\left(2\sin\frac{x}{2}\right)^{2m}.$$

On the other hand, application of the binomial theorem gives

$$(1-e^{-ix})^{2m} = \sum_{j=0}^{2m} (-1)^j \binom{2m}{j} e^{-ijx}. \tag{30}$$

From equations (29) and (30)

$$\sum_{j=0}^{2m} (-1)^{j-m} \binom{2m}{j} e^{-i(j-m)x} = \left(2\sin\frac{x}{2}\right)^{2m} \tag{31}$$

can be derived. Finally, changing the index j to $k = j - m$ yields the desired result.

References

[1] E. T. WHITTAKER, *On a new method of graduation*, Proc. Edinborough Math. Soc. **41**, 63-75 (1923).

[2] O. L. MANGASARIAN, L. L. SCHUMAKER, *Discrete splines via mathematical programming*, SIAM J. Control **9**, 174-183 (1971).

[3] P. H. ASTOR, C. S. DURIS, *Discrete L-Splines*, Numer. Math. **22**, 393-402 (1974).

[4] M. KRYSTEK, *Discrete L-spline filtering in roundness measurements*, Measurement **18**, 129-138 (1996).

Advanced Mathematical Tools in Metrology III
Edited by P. Ciarlini, M.G. Cox, F. Pavese & D. Richter

AN APPROXIMATION METHOD FOR THE LINEARIZATION OF TRIDIMENSIONAL METROLOGY PROBLEMS

L. MATHIEU, C. LARTIGUE, P.BOURDET
Laboratoire Universitaire de Recherche en Production Automatisée
Ecole Normale Supérieure de Cachan, 61 Avenue du Président Wilson
94235 Cachan cedex - France
lartigue@lurpa.ens-cachan.fr

In a previous work we have shown how the Small Displacement Torsor (SDT) concept leads to express the standardized tolerances in a general constrained optimization problem. The deviations between measured points and the geometrical element to identify, as well as the constraints between elements can be linearized at the first order relative to the small displacements. Although the expression of equations using the SDT concept is efficient in most cases, it presents some difficulties for the identification of cylinders the radius value of which is of the same order of the small displacements (straightness of an axis for example). The linearization of the deviation is no more possible. So the problem can be treated as a non linear one. In the paper, we present a solution that allows to completely treat the problem in a linear form. A comparison with classical methods is then made.

1 Introduction

The main objective of tridimensional metrology is to verify the conformity of mechanical parts as regards the standardized tolerances, using data issued from the CMM.

In a previous work[1] we have shown how the Small Displacement Torsor (SDT) concept leads to express the standardized tolerances in a general constrained (or not) optimization problem. The deviations (e_i) between the measured points and the geometrical element to identify (plane, sphere, cylinder, ...), as well as the constraints between elements (fixed distance, given orientation,....) can be linearized at the first order, relative to the components of the SDT. The mathematical resolution method is linear, in the case of the least-squares criterion, or it calls for linear programming techniques (simplex), in the case of other criteria[2]. Although the expression of equations using SDT is efficient in most cases, it presents some difficulties for the identification of elements which are characterized by intrinsic dimensions as the radius (cylinder, sphere, circle), in particular when the radius value is small (treatment of the straightness of an axis, location of pattern of holes, ...). In fact, when the radius value is of the same order as the small displacements value, the linearization is not possible. In most cases, the problem can be treated as a non linear one.

Our purpose is to present an approach that allows to express the treatment of the standardized tolerances in a linear form. This is performed by the approximation of the curvilinear shapes by polyhedral shapes. For example, a cylindrical shape is approximated by a set of couples of parallel planes. Then, the deviations (e_i) are expressed between the measured points and the approximated element. Through the example of the identification of a cylindrical element (straightness of an axis), we present how the approximation leads to a linear optimization problem that can be solved easily by linear programming techniques. The approach can obviously be applied to all kinds of tolerances.

2 The SDT and the elements of small intrinsic dimensions

In the set of standardized tolerances characterizing a mechanical part, an identification problem that frequently appears is that of the identification of the minimum circumscribed cylinder to a set of data points.

The particularity of such a problem is the small distance of the points that characterize the toleranced element, to the cylinder axis that defines the tolerance zone. This is the particular case of straightness of a median line and location of a pattern of holes (fig. 1).

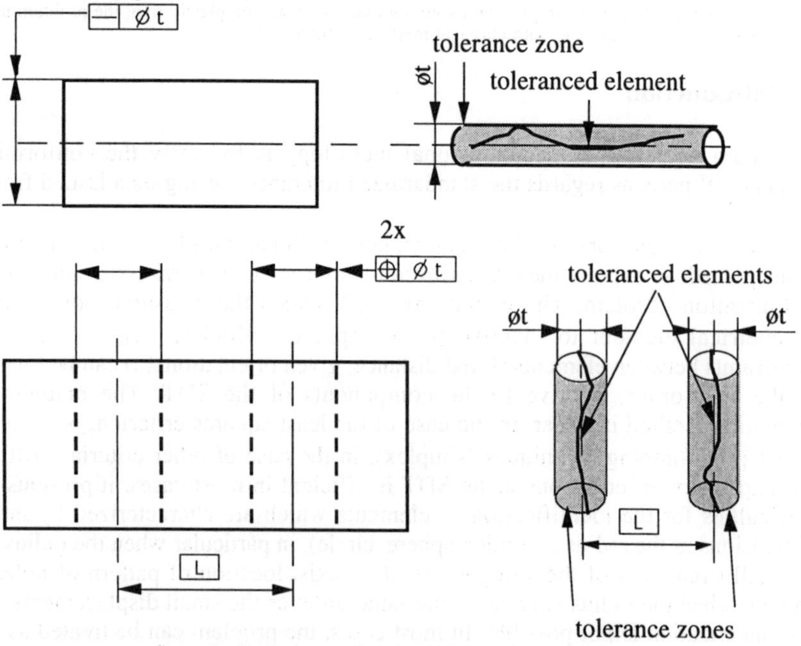

figure 1 : Straigthness of a median line and location of holes

The search for the minimum circumscribed cylinder to a set of n data points consists in identifying the smallest radius cylinder that encloses all the points. The formulation of such a problem is in general a nonlinear constrained minimization and can be expressed as follows.

Let d_i be the distance between each measured point M_i and the axis of the cylinder CY, the radius of which is ρ.

d_i is defined by Eq. 1, where $\vec{t}$ represents a vector director of the axis and A, one point on the axis (fig. 2).

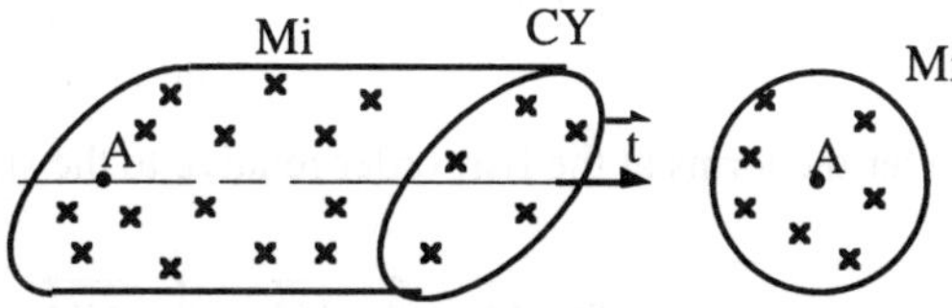

figure 2

$$d_i = \frac{\| \overrightarrow{AM_i} \wedge \vec{t} \|}{\|\vec{t}\|} \quad (1)$$

The problem is : $\min \rho$
subject to $d_i - \rho \leq 0, \quad i \in [1,n]$ (2)

The problem defined by Eq. 2 is, in general, not solved in its nonlinear form. Anthony and al[2] treat the problem in 3 steps: projection of data points in a plane, search for the minimum circumscribed circle in that plane, and adjustment by iterations of the direction for which the cylinder radius is minimal. In most cases a linearization of the equations is proposed[1,3] and this can be carried out using the SDT concept. Consider an initial position of the cylinder close to the solution, $CY'(A', \vec{t}', \rho')$.

The small displacement of CY' involves the A' displacement in A, and the $\vec{t}'$ transformation in $\vec{t}$ and therefore : $\overrightarrow{D_{A'}} = \overrightarrow{A'A}$ and $\vec{t} = \vec{t}' - \vec{t}' \wedge \vec{R}$ (fig.3).

The problem is to find the SDT components $\{\vec{R}, \overrightarrow{D_{A'}}\}$ and the radius ρ that allow to reach the solution.

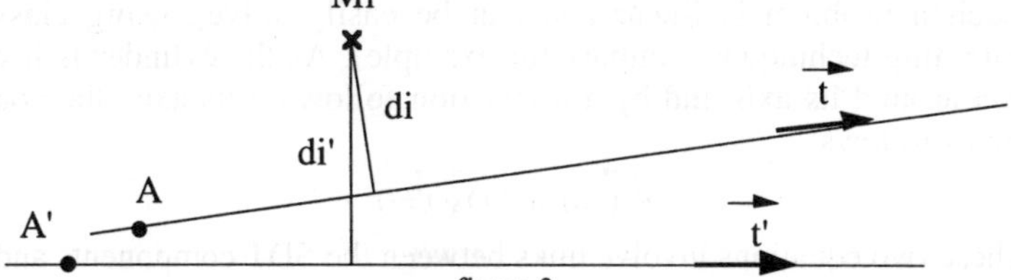

figure 3

Let us now express the distance d_i using the SDT components :

$$d_i = \frac{\| (\overrightarrow{AA'} + \overrightarrow{A'M_i}) \wedge (\vec{t'} - \vec{t'} \wedge \vec{R}) \|}{\| \vec{t'} - \vec{t'} \wedge \vec{R} \|} \quad (3)$$

$$d_i = \frac{\| (\overrightarrow{A'M_i} \wedge \vec{t'} - \overrightarrow{A'A} \wedge \vec{t'} - \overrightarrow{A'M_i} \wedge (\vec{t'} \wedge \vec{R}) + \overrightarrow{A'A} \wedge (\vec{t'} \wedge \vec{R}) \|}{\| \vec{t'} - \vec{t'} \wedge \vec{R} \|}$$

If we only consider the terms at the fisrt order relative to the small displacements :

$$d_i = \frac{\| (\overrightarrow{A'M_i} \wedge \vec{t'} - \overrightarrow{A'A} \wedge \vec{t'} - \overrightarrow{A'M_i} \wedge (\vec{t'} \wedge \vec{R}) \|}{\| \vec{t'} - \vec{t'} \wedge \vec{R} \|}$$

In order to linearize this expression, we introduce a vector $\vec{a_i}' = \dfrac{\overrightarrow{A'M_i} \wedge \vec{t'}}{\| \overrightarrow{A'M_i} \wedge \vec{t'} \|}$

A projection is made on this vector and on the orthogonal vector and then :

$$\| \vec{t'} - \vec{t'} \wedge \vec{R} \|^2 d_i^2 = \left(\| \overrightarrow{A'M_i} \wedge \vec{t'} \| \vec{a_i}' - ((\vec{T}_{A'} \wedge \vec{t'}).\vec{a_i}') \vec{a_i}' + ((\overrightarrow{A'M_i}.\vec{t'}))(\vec{R}.\vec{a_i}') \vec{a_i}' \right)^2$$

$$+ \| -\vec{a_i}' \wedge ((\vec{T}_{A'} \wedge \vec{t'}) \wedge \vec{a_i}') - \vec{a_i}' \wedge (((\overrightarrow{A'M_i}.\vec{R})\vec{t'} - (\overrightarrow{A'M_i}.\vec{t'})\vec{R})) \wedge \vec{a_i}') \|^2$$

The hypothesis of small displacements involves that only the component on $\vec{a_i}'$ is taken into account :

$$d_{iL} = \| \overrightarrow{A'M_i} \wedge \vec{t'} \| - (\vec{T}_{A'} \wedge \vec{t'}).\vec{a_i}' + (\overrightarrow{A'M_i}.\vec{t'})(\vec{R}.\vec{a_i}') \quad (4)$$

In the Eq. 4 d_i has a linear expression relative to the SDT components, and the problem defined by Eq. 2 can be expressed as follows:

$$\begin{aligned} &\min \rho \\ &\text{subject to } d_i - \rho \leq 0, \; i \in [1,n] \end{aligned} \quad (5)$$

where d_i is defined by Eq. 4.

Such a problem is linear and can be easily solved using classical linear programming techniques (simplex for example). As the cylinder is invariant by a rotation around its axis and by a translation following its axis, the conditions are written as follows:

$$\vec{R}.\vec{t'} = 0 \text{ and } \vec{D}_{A'}.\vec{t'} = 0 \quad (6)$$

These two equations involve links between the SDT components and reduce the number of unknowns.

In order to apply this problem to cylinders with great and small radius values, we use the data given by Huang and al[4] and used by Carr and Ferreira. Table 1 shows the results obtained for the two sets of data.

table 1 : Identification of the Minimum Circumscribed Cylinder

Radius using	SDT concept	Carr and Ferreira[3]
Dataset 1	60,057881	60,057881
Dataset 2	*not sufficiently constrained*	0,450055

It can be seen that for the case of the small diameter (above 0.45) no solution is found; *the problem is not sufficiently constrained.*
In fact, for this kind of case, the radius value to find is of the same order as the small displacements and the linearization is no more possible. The linearized distances to the axis are, in reality, the expression of the deviations between points and the cylinder axis projected on the $\vec{a_i}'$ axis. As the radius is of the same order as the displacements, the value of the deviations can become negative.

If the problem defined by Eq. 4 is treated adding the constraints that all the deviations are positive ($d_i \geq 0$), a solution is found but extremely dependent on the initialization and to the number and repartition of the measured points.

Carr and Ferreira give a solution to this problem by limiting the evolution of the SDT components, the limits being adjusted in order to reach the solution. Huang and al only treat the problem of bidirectional tolerance zones. But we can notice that the problem expressed by Eq. 2 can be solved by a non linear solver.

3 The approximation method

Our purpose is to propose a general formulation in a linear form for all the kinds of standardized tolerances. As previously exposed and whatever the identification to carry out, we have to cleverly constrain the problem more, which means disregarding the measured points .

We impose a local *facetization* of the shape, i.e. in each measured point the domain is constrained by couples of parallel lines (fig. 4). This corresponds to a local approximation of the curvilinear shape by a line.

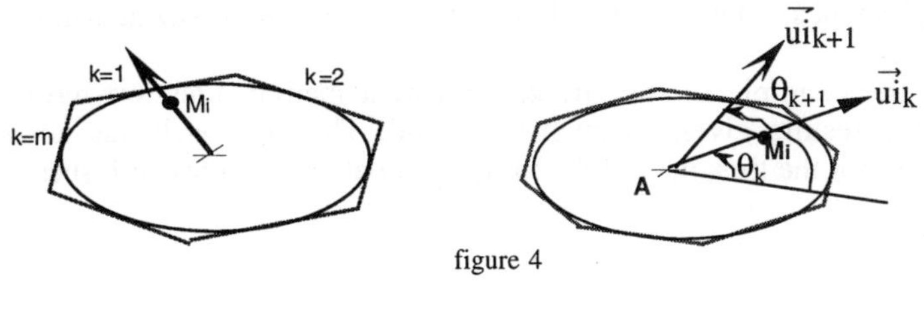

figure 4

Then for each point, we project the vector $\overrightarrow{a_i}'$ on the normal to the each line that defines the polygon, $\overrightarrow{u_{ik}}$ (k varying from 1 to m, where m is the number of polygonal sides).

In the case of the straightness of a median line, the problem can be expressed as follows:

$$\begin{aligned} &\min \rho \\ &\text{subject to } |e_{ik}| - \rho \le 0,\ i \in [1,n] \text{ and } k \in [1,m] \end{aligned} \tag{7}$$

with $e_{ik} = d_{iL}\ \overrightarrow{a_i}'.\overrightarrow{u_{ik}}$

This new linear constrained problem is tested through the same dataset points. The results are given in table 2 and are compared with those given by Carr and Ferreira. The last column corresponds to the results obtained by the non linear problem defined by Eq. 2.

table 2: Identification of the Minimum Circumscribed Cylinder

Radius using	SDT and facetization	Carr and Ferreira[3]	nonlinear
Dataset 1	m= 1; 60,057881	60,057881	60,057881
Dataset 2	m= 120; 0,449806	0,450055	0,449973

Table 2 shows the efficiency of the proposed approach : the values found are very close to those obtained with the nonlinear method and rather better than those given by Carr and Ferreira. *Great cylinders* as well as *small cylinders* can be treated in the same manner. The algorithm works two ways : whether with a great number of facets and one iteration or with a small number of facets and a few iterations.

This approximation can obviously be applied to all tolerances and allows to maintain a formalism based upon the SDT concept.

4 Conclusion

In the paper, we have proposed an approach that allows to express the problem of the treatment of standardized tolerances in a linear form easy to carry out. The formulation is general and gives an answer to the treatment of all the strandardized tolerances. The basis of the approach is the approximation of curvilinear shapes by polygonal shapes. The writing of the problem is still realized using the SDT components.

Through the example of the straightness of a median line, we have seen the satisfactory results given by such a method. This approach has already been carrying out for the treatment of the location of pattern of holes with success[5].

References

1. Bourdet P. and al, The concept of the Small Displacement Torsor in Metrology, *Advanced Mathematical Tools in Metrology II,* Series on advances in mathematics for applied science, vol.40, pp 110-122, (World Scientific, Singapore, 1996)

2. Anthony G.T. and al, Chebyshev best-fit geometric elements, *NPL REPT DITC 221/93,* (National Physics Laboratory, UK, Sept. 1993)

3. Carr K. and Ferreira P., Verification of form tolerances. Part II: Cylindricity and straightness of a median line, *Precision Engineering,* vol.17, n°2, pp 144-156, Avril 1995

4. Huang S.T. and al, A new minimum zone method for evaluating straightness errors, *Precision Engineering,* vol.15, n°3, pp 158-164, July 1995

5. Bourdet P. and al, Mathematical method for tridimensional verification of tolerances of location with degrees of freedom, *Proceedings of the XIII IMEKO World Congress* - vol.3, PP 1641-1645, Torino, September 1994

Advanced Mathematical Tools in Metrology III
Edited by P. Ciarlini, M.G. Cox, F. Pavese & D. Richter

REGULARIZATION ALGORITHMS FOR IMAGE RECONSTRUCTION FROM PROJECTIONS

E.LOLI PICCOLOMINI, F.ZAMA
Dept. of Mathematics, University of Bologna,
Piazza di Porta S. Donato 5, 40127 BOLOGNA (Italy)

There is a great variety of methods for computing tomographic reconstructions. In our work, we consider the discretized problem when a limited number of projections is available, and we search for a regularized least squares solution of the undetermined projections system. The results obtained using the Tikhonov method, the LSQR and the CGNR methods, are reported.

1 Introduction

The problem of image reconstruction refers to the recovery of values of 3D functions from cross–sectional projections. The solution of this problem has been widely treated in literature[1] and a lot of methods have been studied[2]. The problem of image reconstruction from projections is usually stated as follows:

$$g(\theta, t) = \int_{-\infty}^{\infty}\int_{-\infty}^{\infty} f(x, y)\delta(t - x\cos(\theta) - y\sin(\theta))dxdy \qquad (1)$$

where δ is an impulse function. The value of $g(\theta, t)$ represents the integral of $f(x, y)$ along the line of orientation $\theta + \pi/2$ at displacement t from the center of the field. In practice, the reconstruction requires finite dimensional representation of both the observations $g(\theta, t)$ and the function $f(x, y)$. To achieve this, we suppose that the domain of $f(x, y)$ is a square and we subdivide it into $n \times n$ cells D_i, $(i = 1, \dots, n^2)$ using a uniformly spaced square grid. In a vector $\mathbf{f}$ we store the mean value of $f(x, y)$ on each cell D_i following the lexicographic arrangement. The observations are discretized by measuring the line projection of the image at angle θ $(0 \le \theta \le \pi/2)$ onto the horizontal: for simplicity n values for each angle are measured. If we consider p projections at angles $\theta_1, \dots, \theta_p$, we can store the observations in a vector g of $p \times n$ components. The projection values are assumed to be linear combinations of the image values and may be described as:

$$g_k = \sum_{j \in S_k} f_j \quad k = 1, \dots, m \;\; (m = n \times p) \qquad (2)$$

where S_k is made up of all cells D_i that contribute to the projection g_k. In matrix notation the projection equations (2) can be written as a set of m linear equations:

$$\mathbf{g} = H\mathbf{f} \tag{3}$$

where H is an $m \times n^2$ matrix:

$$h_{k,j} = \begin{cases} 1 & \text{if } j \in S_k \\ 0 & \text{otherwise} \end{cases} \tag{4}$$

Since usually $p < n$, (3) is an undetermined linear system with a rank deficient matrix. In the previous formulation, we have assumed that the projections are exactly measurable; in practical applications, however, we can only measure the values $\mathbf{z} = \mathbf{g} + \eta$, where the last term represents the noise (usually Gaussian white noise). One apprach to solve (3) consists in finding a solution to the least squares problem:

$$min \parallel H\mathbf{f} - \mathbf{z} \parallel_2 \tag{5}$$

The problem is usually ill–conditioned and it is necessary to solve it by regularization methods. We analyze direct and iterative methods for an efficient computation of the regularized least squares solutions of (3). In particular in section 2 we briefly analyze the Truncated Singular Value Decomposition method. Then in section 3 we introduce the Tikhonov regularization method and its computational implementation. Finally, in section 4 the conjugate gradient type methods are reported. The computational experiments discussed in section 5 are carried out on a Sun Spark 20 with Fortran and C codes.

2 The TSVD method

The Singular Value Decomposition (SVD) is of great importance for problem(5). The solution can be explicitly written as: $x_{LSQ} = \sum_{i=1}^{r} \frac{\mathbf{u}_i^t \mathbf{z}}{\sigma_i} \mathbf{v}_i$ where $r = rank(H)$, $\sigma_1 \geq \sigma_2 \geq, \ldots \geq \sigma_r > 0$ are the singular values and $H = \sum_{i=1}^{r} \mathbf{u}_i \sigma_i \mathbf{v}_i^t$ ($r \leq m$ and $U = (\mathbf{u}_1, \ldots, \mathbf{u}_m)$, $V = (\mathbf{v}_1, \ldots, \mathbf{v}_n)$ are orthogonal matrices). The SVD of the matrix H gives additional information about the ill–conditioning of the matrix through the condition number σ_1/σ_r and the Fourier coefficients $\mathbf{u}_i^t \mathbf{z}$ give information about the quality of the least squares solution. If $\mathbf{u}_i^t \mathbf{z}$ do not decay faster than σ_i then the solutions are dominated by the components relative to the smaller values σ_i and the solution may appear as completely random. The purpose of regularization methods is to filter out the contribution to the solution corresponding to the small singular values. Then the regularized solution $\mathbf{f}_{reg}$ can be expressed in terms of the *filter factors* $\phi(\sigma_i)$

$$\mathbf{f}_{reg} = \sum_{i=1}^{r} \phi(\sigma_i) \frac{\mathbf{u}_i^t \mathbf{z}}{\sigma_i} \mathbf{v}_i \tag{6}$$

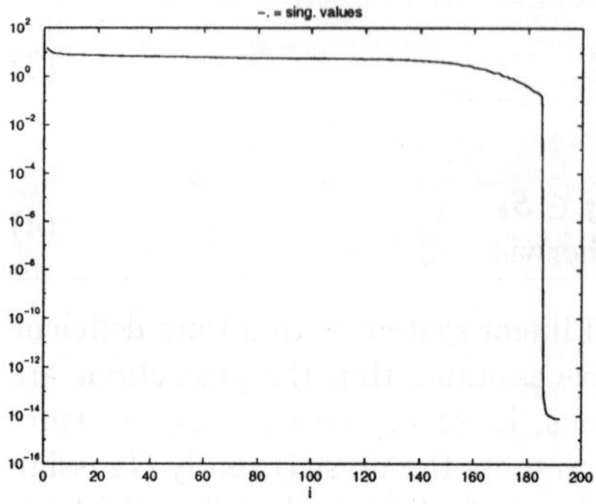

Figure 1: Singular values in Test problem I1: 6 angles

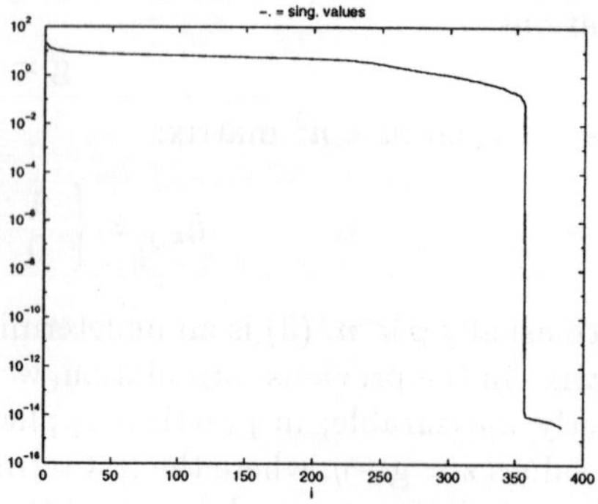

Figure 2: Singular values in Test problem I1: 12 angles

The choice of the *filter factors* is characteristic for the particular regularization method. The Truncated SVD (TSVD) method [5] solves a new problem derived from (5) with a better conditioned matrix H_K of rank $K < r$: $H_K = \sum_{i=1}^{K} \mathbf{u}_i \sigma_i \mathbf{v}_i^t$ and the solution of the problem $min \parallel H_K \mathbf{f}_K - \mathbf{z} \parallel^2$ can be written as [6]: $\mathbf{f}_K = \sum_{i=1}^{K}((\mathbf{u}_i^t \mathbf{z})/\sigma_i)\mathbf{v}_i$.

3 Tikhonov Method

Another highly regarded method for solving (5) is the Tikhonov regularization method, which corresponds to the solution of: $min \parallel H\mathbf{f} - \mathbf{z} \parallel^2 + \lambda \parallel Q\mathbf{f} \parallel^2$ where Q is is a suitable *regularization matrix.* The regularized solution $\mathbf{f}_\lambda$ is then obtained by the normal equations:

$$(H^t H + \lambda Q^t Q)\mathbf{f}_\lambda = H^t \mathbf{z} \tag{7}$$

In practical applications the regularization matrix Q is the discretization of some differential operator which expresses the smoothness of the solution. One choice commonly used in tomographic reconstruction [3] is to define the regularization matrix by means of the discrepancy between the actual image and the reconstructed picture:

$$\Phi(\mathbf{f}) = \frac{1}{2}\mathbf{f}(I + \alpha C^t C)\mathbf{f} \tag{8}$$

where $\mathbf{y} = C\mathbf{f}$ has the following expression:

$$\mathrm{y_j} = \mathrm{f_j} - \tfrac{1}{8}(\mathrm{f_{j-n-1}} + \mathrm{f_{j-n}} + \mathrm{f_{j-n+1}} + \mathrm{f_{j-1}} + \mathrm{f_{j+1}} + \mathrm{f_{j+n-1}} + \mathrm{f_{j+n}} + \mathrm{f_{j+n+1}})$$

and α is a positive weight factor. Since the matrix $(I + \alpha C^t C)$ results positive definite we can determine Q such that: $Q^t Q = (I + \alpha C^t C)$. The common approach to the numeric solution of (7) with direct methods requires $O(N^3)$ operations ($N = n^2$). In the solution of 2D and 3D problems this value can be too large and better results can be obtained by exploiting the structure of the matrices H and $I + \alpha C^t C$. Since the matrix of the normal equations (7) is positive definite, the conjugate gradient method can be suitably used. The number of nonzero elements on each row of H does not exceed $3n$ and the matrix C has at most 9 non zero elements per row. The computational complexity depends on the the number of iterations required and on the computational cost of the matrix vector products $y = Hx$ ($y = H^t x$) and $y = Cx$ ($y = C^t x$). In our application, they all require O(N) operations. The computation of the optimal regularization parameter λ^* usually requires the solutions $\mathbf{f}_\lambda$ for different values of λ. However, in this particular application, a good regularization parameter is given by: $\tilde{\lambda} = 1./(max(diag\{H^t H\}))$. Suitable preconditioning techniques can be applied to the conjugate gradient method in order to reduce the number of iterations. The following preconditioning matrix has been used in our numerical experiments: $\Delta = diag\{H^T H + \lambda Q^t Q\}$.

4 Iterative methods

An alternative approach to the regularization problem is represented by the iterative methods applied to the system of normal equations obtained from the problem (5). In particular, we consider the conjugate gradient method (CGNR) and the LSQR method of Paige and Saunders[10].

Some regularization properties of these methods have been analyzed in the last years [12,11]. They can be derived in terms of the SVD expansion of the iterates and by examining the filter factors obtained. When the number of iterations exceeds a threshold value, the solution diverges from the regularized one and the method must be stopped. Hence the number of iterations plays the role of regularization parameter for the method. It is evident that the method is efficient only if a good solution is obtained with few iterations and this generally depends on the behaviour of the singular values of H.

A technique proposed by Hanke and Hansen [8] to solve a better conditioned problem is that of preconditioning the system of the normal equations by means of the inverse regularization matrix. In our work we implement this iterative method (PCGNR) using the Cholesky factorization of the regularization matrix $I + \alpha C^t C$ introduced in (3). Each iteration requires therefore the solution of two triangular linear systems. The success of CGNR and PCGNR mainly depends on the stopping rule adopted for the algorithm. There are several proposals in literature [8]: we have utilized a criterion that considers both the residual and the Euclidean norm of the solution, as in all the regularization

Table 1: (random white noise variance 8.2e-2)

TEST	Method	λ	PE(> 70)	NRMSE	NMAE	Iter	sec
I1	TIKH	5.e-2	10.1%	0.6	0.3	44	0.39
	CGNR		2.83%	0.5	0.25	5	4.5(-2)
	PCGNR		9.3%	0.56	0.3	5	6.36
	LSQR		2.83%	0.5	0.25	5	7.8(-2)
	TSVD	(50)	15.2%	0.75	0.38		1.16
I2	TIKH	1.8(-2)	2.1%	0.38	0.19	159	46
	CGNR		2.5%	0.40	0.21	50	14.33
	PCGNR		2.2%	0.40	0.20	50	203
	LSQR		2.4%	0.40	0.21	50	35.82
I3	TIKH	9.(-3)	1.96%	0.35	0.17	343	597
	CGNR		2.94%	0.39	0.19	59	126
	LSQR		3.46%	0.41	0.2	50	98.55
I4	TIKH	4.0(-3)	0.01%	0.18	0.09	146	53
	CGNR		0.03%	0.19	0.1	50	19.44
	PCGNR		0.02%	0.17	0.08	50	1005
	LSQR		0.03%	0.19	0.1	50	29.1

methods. In order to complete the discussion on the conjugate gradient method as a regularization method, we underline some computational features. The system coefficient matrix need not be computed, because the cross product $(H^tH)x$ is executed by two successive matrix–vector products with computational complexity of $O(N)$: this reduces the computational complexity and takes care of the sparsity of H in our case.

An equivalent formulation of CGNR is given by the LSQR method[8] applied to the normal equations, where the implicit and explicit execution of the cross product H^tH are avoided. In the exact arithmetic the iterates x_k^{CG} and x_k^{LSQR} are exactly the same, but in the finite arithmetic the LSQR algorithm should be more stable.

5 Numerical Experiments

In this section we show some results obtained with the methods previously introduced. Results that we report have been carried out on four test problems: I1 (brain section image of 32 × 32 pixels with 6 projection angles), I2 (brain section image of 64 × 64 pixels with 40 projection angles), I3 (brain section image of 128 × 128 pixels with 80 projection angles), I4 (Rx image of 90 × 90 pixels with 25 projection angles). The pictures $\tilde{\mathbf{f}}$ reconstructed with different methods have been compared with the original images **f** by means of the Herman error parameters:

$$NRMSE = \sqrt{\frac{\sum_{v=1}^{n}\sum_{u=1}^{n}(\mathbf{f}_{u,v} - \tilde{\mathbf{f}}_{u,v})^2}{\sum_{v=1}^{n}\sum_{u=1}^{n}(\mathbf{f}_{u,v} - \overline{\mathbf{f}})^2}} \quad NMAE = \frac{\sum_{v=1}^{n}\sum_{u=1}^{n}|\mathbf{f}_{u,v} - \tilde{\mathbf{f}}_{u,v}|}{\sum_{v=1}^{n}\sum_{u=1}^{n}|\mathbf{f}_{u,v}|}$$

where $\overline{\mathbf{f}}$ is the average of the values in original image **f**. A further information on the quality of the reconstructed images is given by the percentage (EP) of points $\tilde{\mathbf{f}}_\mathbf{i}$ such that $|\mathbf{f}_\mathbf{i} - \tilde{\mathbf{f}}_\mathbf{i}| > l$ (we have chosen $l = 70$). In table 1 we report the errors (EP, NMRSE, NMAE), the iteration number (Iter) and the execution time (sec) obtained using the Tikhonov method (TIKH), the Conjugate Gradient Method (CGNR), the Preconditioned Conjugate Gradient Method (PCGNR) and the LSQR method. The regularization parameter $\tilde{\lambda}$ reported in correspondence of the TIKH method is obtained as described in section 3; anyway, for the test problems discussed, a wide range of values ($< \tilde{\lambda}$) of the regularization parameter give good solutions.

By looking at table 1, the first consideration is that in each test problem the errors obtained with all the methods presented are very similar. In terms of computational times, the CGNR method is obviously the fastest and hence it can be globally considered the more efficient for this kind of problems. The behaviour of the CGNR method can be related to that of the singular values of H [8], in the sense that a slow decreasing of the singular values corresponds to the decreasing of the solution error in the iterations of the CGNR; a gap in the curve of the singular values usually reflects in a strong oscillation of the CGNR solution. If we look at figure 1, relative to I1, we can easily guess that only in the final iterations the CGNR solution should become rather unstable and hence it is possible to stop the method when a good approximation has been already obtained. It is not possible to report the SVD for the larger cases I2, I3 and I4, for problems of memory space, but the results of the methods presented in table 1 let us suppose that the previous considerations can be extended even for these cases. If we compare the solutions of the iterative methods with that of TIKH method, we notice that the regularization doesn't lead to better results in general. Moreover, the partial solutions obtained when solving the system (7) with conjugate gradient are very similar to those of CGNR method. If the test problem is further perturbed with random white noise with variance in the interval (0,1], the numerical results are essentially the same. This brief discussion leads to the conclusion that the problem of

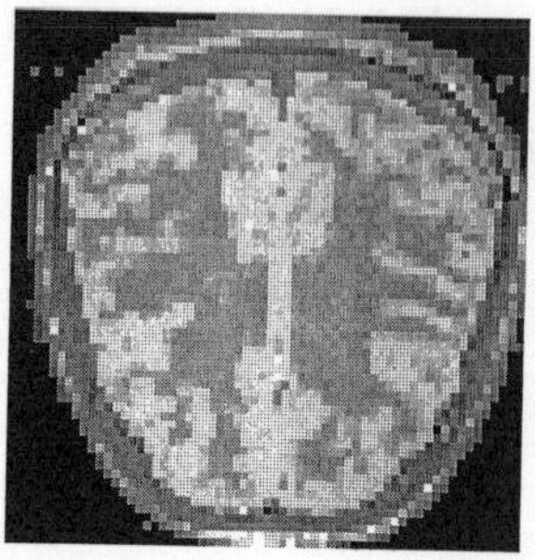

Figure 3: Test image I3

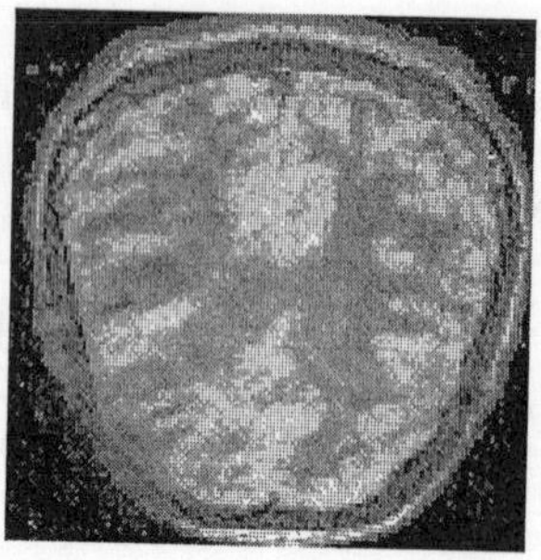

Figure 4: Reconstruction with Tikhonov method.

image reconstruction from projections can be classified as a middle ill–posed problem; it can be efficiently solved with an iterative method, such as conjugate gradients, applied to the normal equations.

References

1. G.T Herman, Image Reconstruction from Projections, (Academic Press, 1980)
2. F. Natterer, The Mathematics of Computerized Tomography, (John Wiley & Sons 1989)
3. R.L Kayshap and C. Mittal, *IEEE Trans. on Computers* **24**, 915 (1975).
4. P.C.Hansen, *Numerical Algorithms* **6**, 1 (1994)
5. J.M.Varah, *SIAM J. Numer. Anal.* **10**, 257 (1973)
6. P.C.Hansen, *SIAM J. Sci. Stat. Comp.* **11**, 503 (1990)
7. P.C.Hansen, *BIT* **27**, 543 (1987)
8. M. Hanke, P.C.Hansen, *Surveys on Mathematics for Industry* **3**, 543 (1993)
9. M. Hanke *Pitman Reserch Notes in Mathematics* , 327 (1995)
10. C. C. Paige, M. A. Saunders,*Anal Comp Math* **8**, 43 (1982)
11. R. Plato, G. Vainikko, *Numer. Funct. Anal. Optim.* **11**, 111 (1990)
12. A. van der Sluis, H. A. Van der Vorst, *Linear Algebra Appl.* **130**, 257 (1990)

Advanced Mathematical Tools in Metrology III
Edited by P. Ciarlini, M.G. Cox, F. Pavese & D. Richter

QUALITY OF EXPERIMENTAL DATA IN HYDRODYNAMIC RESEARCH

M. MASIA, R. PENNA
Istituto Nazionale per Studi ed Esperienze di Architettura Navale, I.N.S.E.A.N., Via di Vallerano 139, 00128 Roma, ITALY

In this paper we would point out the adopted methodology to achieve the uncertainty in Experimental Fluid Dynamics (EFD). A brief description of the Uncertainty Analysis and its application to recent experimental results are reported.

1 Introduction

The development of marine vehicle has been an evolutionary process in which engineering progress in the naval community was generally based on previous experience and databases of experimental data. Nowadays, Computational Fluid Dynamics (CFD) applications are becoming more and more complex and the availability of experimental data obtained with accuracy requirements and Uncertainty Analysis becomes more and more important, because CFD validation requires an effective and accurate experimental activity. On the other hand numerical results to be validated should satisfy suitable quality requirements. This activity implies a close co-operation between researchers working both in numerical and experimental area.

The typical experimental facility to carry out investigation about fluid dynamics of marine vehicles is the Towing Tank. That is a very large tank some hundreds meters long, with a carriage, said dynamometric carriage, running at controllable speed. The ship model equipped by the required devices is guided or constrained to keep the steady carriage speed, that is the scaled speed, according to Froude law, of the designing ship. A typical experiment is carried out through some runs in a range of Froude number and can achieve global or local measurements. Total resistance, sinkage, trim, sideforces, propeller efficiency are the most of global measurements, whereas local measurements regard the flow investigation due to particular hull shape.

2 Overview of the methodology

By definition, no measurement is perfectly accurate, therefore the goal of Uncertainty Analysis (UA) is an assessment of the difference between the measured value and the true value. This error in measurement is composed of bias and precision components. The bias error is the fixed and systematic component of the total error that has the same value for each measurement. The precision error is the random component of the total error and is the measure of the repeatability and stability of a measurement. Bias can only be estimated and can be reduced by calibration, whereas the precision error is determined by taking N readings of a parameter X and computing the precision index (PI) from

$$S_X = \left[\frac{1}{N-1} \sum_{i=1}^{N} (X_i - \overline{X})^2 \right]^{1/2} \tag{1}$$

where $\overline{X}$ is the average of X. The PI of the mean of the N readings is:

$$S_{\overline{X}} = \frac{S_X}{\sqrt{N}} \tag{2}$$

Every error source has either or both bias and precision components. The total bias and precision error are combined with the root-sum-square method (RSS), such as, respectively:

$$B = \left[B_1^2 + B_2^2 + ... + B_k^2\right]^{1/2} \tag{3}$$

$$S = \left[S_1^2 + S_2^2 + ... + S_k^2\right]^{1/2} \tag{4}$$

Precision limit (P) is defined as $P = tS$ where t is the Student's distribution coefficient. Finally the total uncertainty (U) is a combination of B and P according to additive method (ADD)

$$U_{ADD} = (B + P) \tag{5}$$

or to RSS method:

$$U_{RSS} = \left(B^2 + P^2\right)^{1/2}. \tag{6}$$

Either (5) or (6) is chosen when 99% or 95% coverage estimate is required.

Nevertheless, each result of interest in an experiment is obtained through a data reduction equation, that reflects the situation that for many results of interest several variables are measured in an experiment and combined to get the desired result. If the result is a function of j parameters, such as:

$$r = f(X_1, X_2, ..., X_j) \tag{7}$$

the bias and precision index of the result are given below, respectively:

$$B_r = \left[\sum_{i=1}^{j} (\xi_i B_{X_i})^2 \right]^{1/2} \tag{8}$$

$$S_r = \left[\sum_{i=1}^{j}(\xi_i S_{X_i})^2\right]^{1/2} \tag{9}$$

where $\xi_i = \frac{\partial r}{\partial X_i}$, and B_{X_i} and S_{X_i} are bias and precision error of X_i. The coefficients ξ_i, known as sensitivity coefficients, mean the influence degree of each error source on global error, and play a considerable role in their combination.

Finally, the uncertainty in the result can be computed according to (5) or to (6) by substituting B with B_r and S with S_r.

3 Description of the experimental equipment

The development of some CFD codes regarding the prediction of the dynamic behaviour of a maneuvering ship required the availability of accurate experimental data about hydrodynamic forces acting on a bare hull model in steady drift motion.

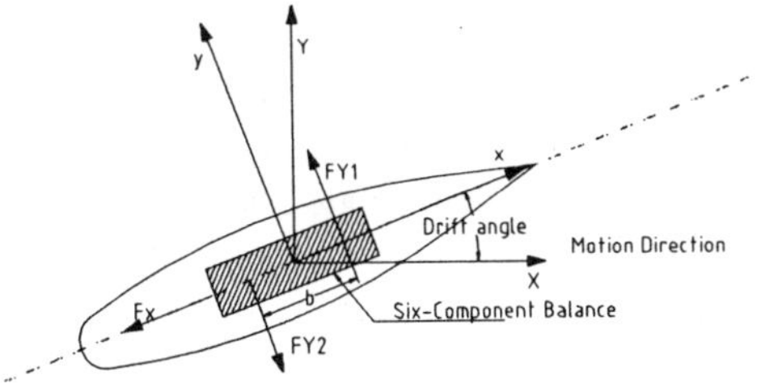

Figure 1: Forces acting on the model in the X-Y plane

The experiment was carried out on a model about six meters long in the INSEAN towing tank n.2. The towing tank is $220m$ long, $9m$ wide and $3.5m$ deep. The forces were measured with a six-components balance, whose lower beam was fastened to the model, and the upper one connected to the carriage by means of a rotating device. In this way, drift angles up to ± 10 deg could be adjusted with an accuracy of ± 0.1 deg. The two beams are connected together with six coupling rods and they can only transmit one force in their longitudinal direction. The measurement of the individual component is made by means of high quality strain gauge force transducers. For the x and y components $500N$ nominal load transducers were used, whereas for the three z oriented components $7300N$ nominal load transducers. The accuracy of the transducers reported by the owner is below $\pm 0.1\%$ of the respective nominal load. To verify this target and to estimate the bias limit of the transducers, some calibration tests were executed with sample weights in the whole measuring range. The calibration results also showed a linear and repeatable behaviour of the transducers and the absence of mutual influence from the components.

The signals were acquired by a 16-channels digital recording with $5KHz$ sampling rate and then transferred to a personal computer resampled with $240Hz$. For each steady condition a $10sec$ measurement period was selected and statistically analyzed.

4 Error sources and their propagation

The kind of experiment carried out is defined as *steady*, in the sense that measures are taken when steady conditions are realized in terms of carriage speed and consequently steady flow around model. In this situation, when we suppose the forces acting on the hull steady and so for the balance transducers, the values can be acquired. Nevertheless to assure the completely steadiness, the entire runs and the *zero-th* levels are recorded and subsequently the steady period selected. Each run is characterized by different carriage speed or drift angle.

The main recognized bias error sources (Tab.1) are located in the model geometry and its main dimensions, the volume displacement, the mechanical arrangement on the carriage, the drift angle setting.

Table 1: Summary of Bias error sources

Model dimensions	$\pm 0.002m$
Drift angle	± 0.1 deg
Calibration: F_x, F_{Y1}, F_{Y2}	$\pm 0.1N$
Calibration: F_{Z1}, F_{Z2}, F_{Z3}	$\pm 1.5N$
A/D Conversion	$\pm 0.0002V/bit$
Velocity	$\pm 0.001m/s$
Water density	$\pm 0.01N/m^3$

Whereas precision errors, at most associated with stability of the load transducers, and steadiness of the carriage towing speed and the flow around the hull, were obtained from multiple readings at two velocities and some drift angle positioning. In order to assure the completely independence of the measurements, the readings were performed in a period some days long. All of the above mentioned bias and precision errors propagate into the final uncertainty according to propagation equations (8,9).

Typical results of this experiment are the nondimensional hydrodynamic sideforces, longitudinal force, vertical forces and moments about the three axis (Fig.1). Data reduction equations, in the case of sideforce and z-moment, are written as:

$$C_y = \frac{F_{Y1} + F_{Y2}}{0.5\rho L T v^2} \tag{10}$$

$$C_{Mz} = \frac{(F_{Y1} - F_{Y2}) b_z}{0.5\rho L^2 T v^2} \tag{11}$$

where F_{Y1},F_{Y2} are the sideforce components, v the steady carriage speed, ρ the water density and L, T respectively the length and draft of the ship model. But the forces are strongly dependent on drift angle β and there is no data reduction equation where this variable is present. In fact, according to (8,9) bias and precision index for sideforce coefficient (10) are:

$$\begin{aligned} B_{Cy} = & \left[\left(\frac{\partial Cy}{\partial F_{Y1}} B_{F_{Y1}} \right)^2 + \left(\frac{\partial Cy}{\partial F_{Y2}} B_{F_{Y2}} \right)^2 + \left(\frac{\partial Cy}{\partial \rho} B_\rho \right)^2 \right. \\ & \left. + \left(\frac{\partial Cy}{\partial L} B_L \right)^2 + \left(\frac{\partial Cy}{\partial L} B_T \right)^2 + \left(\frac{\partial Cy}{\partial v} B_v \right)^2 \right]^{1/2} \end{aligned} \tag{12}$$

$$S_{Cy} = \left[\left(\frac{\partial Cy}{\partial F_{Y1}} S_{F_{Y1}} \right)^2 + \left(\frac{\partial Cy}{\partial F_{Y2}} S_{F_{Y2}} \right)^2 + \left(\frac{\partial Cy}{\partial v} S_v \right)^2 \right]^{1/2} \tag{13}$$

Therefore bias error due to drift angle positioning can not be considered in the final results, unless many readings were taken with many drift angle positioning. Multiple readings are also helpful to highlight different steady flow conditions at hypothetical same initial conditions.

5 Repeatability test results

Due to experimental cost reduction, multiple readings (not less than 20) have carried out at two velocities (low and high Fr) and four drift angles. The test was organized to repeat the readings every four runs, changing drift angle and realizing two steady velocities each run. In this way the complete experiment has been performed with 80 runs during a period seven days long. The distribution of F_{Y1} and F_{Y2} is reported in Fig.2 in the case of high velocity and 2.5 deg of drift angle, that is the worst of the whole experiment.

The above distribution shows a scattering of these forces higher than bias limit reported in Tab.1. This can be ascribed to the effects of drift angle bias or to different steady hydrodynamic condition, due to vortex shedding. In

Table 2: Overall Uncertainty (%) in results (95% coverage estimate)

Hydrodynamic coefficient	Fr	Drift angle			
		2.5 deg	5.0 deg	7.5 deg	10.0 deg
Cx	Low Fr	±1.2	±1.1	±1.2	±1.2
	High Fr	±0.7	±0.7	±0.7	±0.8
Cy	Low Fr	±13.0	±4.2	±2.4	±2.1
	High Fr	±12.3	±3.6	±2.4	±3.7
CMz	Low Fr	±2.9	±1.2	±0.9	±1.0
	High Fr	±2.9	±1.1	±0.9	±1.3

the other conditions forces showed very narrow scatter. In fact, as reported in Tab.2, the global Cy uncertainty, obtained according to RSS method at 95% confidence limit, reflects the behaviour of precision index. It is also interesting to note how same distribution of F_{Y1} and F_{Y2} produces very different uncertainty errors in results (Fig.3) due to different combination by sensitivity coefficients. This is the case for *z-moment CMz* coefficient, whose measured force components are the same ones as for sideforce coefficient Cy, but uncertainty is lower than Cy one.

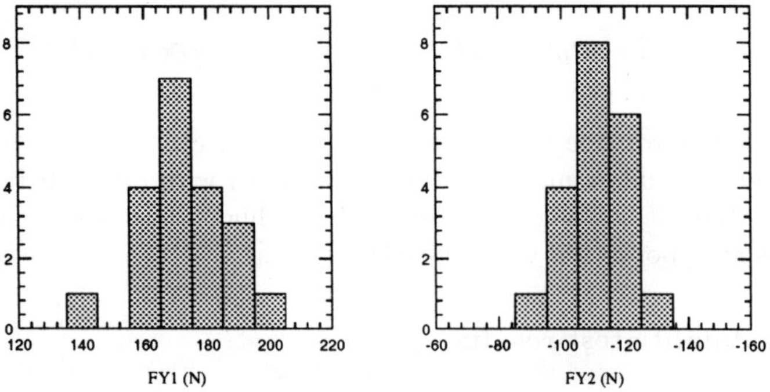

Figure 2: Distribution of F_{Y1} and F_{Y2} readings ($Fr = 0.316$, $\beta = 2.5$ deg).

6 Conclusions and future work

Towing Tanks should spend more effort in the activity to improve the quality of experimental data,by extending Uncertainty Analysis to every kind of experiment. The presented test case is a very simple one, in fact most experiments require complex mathematical formulation to obtain final results and propagation equations become consequently more complex. Moreover, some experiments are performed in non-steady condition, e.g. seakeeping tests, and this methodology could fail, because in this case signals are analysed in time domain. Standards and guidelines are also needed to design a high accuracy experiment.

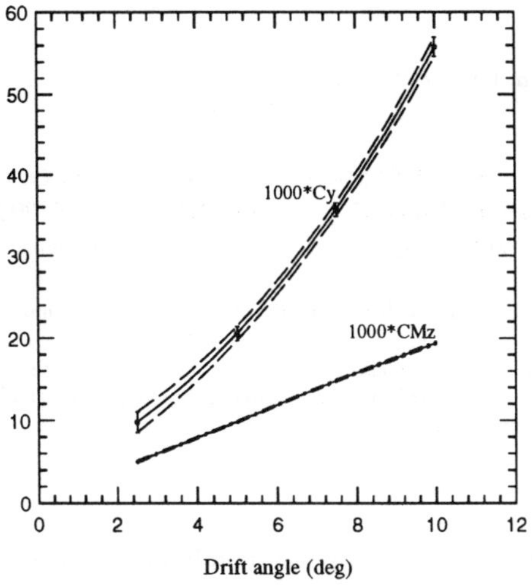

Figure 3: Cy and CMz global uncertainty with 95% confidence limit

Acknowledgements

This work has been performed in the frame of the INSEAN Research Plan 1991-93 supported by the Ministry of Transportation and Navigation with law n.132/94.

The authors are grateful to M. Fortunati for the mechanical design, and to M. Palini and A. Sabelli for the considerable work performed in setting and carrying out the experiments.

References

1. H. W. Coleman, W. G. Steele, *Experimentation and Uncertainty Analysis for Engineers*, ed. Wiley Interscience (1989).
2. *Measurement Uncertainty*, ANSI/ASME PTC 19.1-1985, Part 1 (1986).

Advanced Mathematical Tools in Metrology III
Edited by P. Ciarlini, M.G. Cox, F. Pavese & D. Richter

STOCHASTIC DRIFT MODELS FOR THE DETERMINATION OF CALIBRATION INTERVALS

P. TAVELLA
Istituto Elettrotecnico Nazionale, Strada delle Cacce 91, 10135 Torino, Italy
tavella@tf.ien.it

A. BOBBIO
Dip. Informatica, Università di Torino, C.so Svizzera 185, 10149 Torino, Italy
bobbio@di.unito.it

S. COSTAMAGNA, A. MONTEFUSCO
MEMC Electronic Materials SpA, Viale Gherzi 31, 28100 Novara, Italy
scostamagna or amontefusco@memc.inet.it

The paper discusses different stochastic models for evaluating optimal calibration intervals in measuring instruments. Calibration condition is assumed to be traced by means of an observable parameter that undergoes a drift stochastic process. The statistics of the first passage time of the drift across a tolerance barrier is studied and the optimal calibration interval is set equal to a suitable percentile of its distribution function.

1. Introduction

The assessment of the correct calibration condition of measuring instruments is a very crucial problem, particularly for a high technology company whose quality requirements are more stringent. The calibration tests require, in fact, to suspend the production process and can represent a major source of process unavailability. If the result of a test shows an out-of-tolerance condition, the production process must be stopped, the instrument readjusted, and the parts produced since the previous positive test must be re-examined. Therefore, the estimation of the proper test interval is an important specification in any quality assurance program, and can be formulated as an optimal trade-off between the process unavailability and the risk of performing incorrect measurements. However, there is a surprising lack of well established and recommended methods in the international standards [1,2] and such recommendations mostly apply to a large class of similar instruments, while the method here described is intended for single key instruments whose demanding use deserves a peculiar study. It is assumed that the calibration condition of a measuring instrument can be monitored by means of one observable parameter and that such parameter, initially set to zero, undergoes a stochastic drift process. The drift process is supposed to be caused by a sequence of discrete random shocks whose amplitude and arrival time are described by random variables of known probability distribution functions. The problem is to estimate the first passage time of the drift process across a certain tolerance barrier. The calibration interval is then set equal to a suitable percentile of the distribution function of this first passage time. In the paper, different stochastic drift models are examined and compared. The distribution functions of the corresponding first passage times (and hence the calibration intervals) are inferred. Preliminary results of the experimental validation of the method are eventually reported.

2. Stochastic drift models and calibration interval

The calibration condition of a given instrument can be determined by measuring the deviation of an observable parameter with respect to a preassigned value assumed as the correct one. The measured deviation undergoes a drift that can be represented by a stochastic model [3, 4]. Let $z(t)$ be the value of the measured deviation at time t. The drift in the parameter value is supposed to be caused by a sequence of discrete random shocks whose amplitude is a stochastic variable x with known cumulative distribution function (cdf) $F_x(x)$. Different hypotheses on the functional dependence of the total drift on the single shock give rise to different stochastic models [4]. Let a be a symmetric bilateral threshold representing the tolerated deviation with respect to the correct value. The region delimited by the barriers $+a$ and $-a$ can be considered as the acceptable band of variation on the measured parameter $z(t)$. The instrument is considered out-of-tolerance when the total deviation $z(t)$ exceeds the level $+a$ or $-a$. Let T be the time at which the drift process reaches the value $\pm a$ for the first time. T is the first passage time of $z(t)$ across the tolerated barrier and is a random variable with cdf $H_T(t) = Pr\{T \leq t\}$. Let the corresponding survival function sf be: $\overline{H}_T(t) = 1 - H_T(t)$. Given an acceptance level α (e.g. α=0.95) the calibration interval can be defined as the value θ such that:

$$H_T(\theta) = 1 - \alpha \tag{1}$$

θ is the $(1-\alpha)$-th percentile of the distribution $H_T(t)$, and hence, the event of incurring a crossing of the tolerance threshold before θ occurs with a risk equal to $(1-\alpha)$. In this paper, we examine two cases: the drift can be due to uncorrelated shocks or to additive shocks. In the first case, the subsequent shocks are uncorrelated and do not have any cumulative effect. Hence, the total drift $z(t)$ is equal to the value of the last shock, and the first passage time across the barrier $\pm a$ is the time at which a single shock with an amplitude greater than a (in absolute value) occurs for the first time.

In the additive model, the effects of the subsequent shocks add up and the total drift $z(t)$ is equal to the sum of the shock amplitudes. The first passage time against the barrier occurs when the sum of the successive shock amplitudes exceeds the threshold $\pm a$. This model corresponds to a random walk, in which each increment x is independent from the previous ones and the cumulative effect is observed.

2.1 Non-cumulative drift

Shocks occur at equispaced intervals Δt and are uncorrelated. The deviation *z(t)* is thus equal to the value of the last shock, i.e. $z(k\,\Delta t) = x_k$. The out-of-tolerance condition is reached only when a single deviation x occurs with amplitude greater than $\pm a$. The probability of surviving the first k intervals is given by:

$$\overline{H}_T(k\Delta t) = \Pr\{T > k\Delta t\} = \Pr\{|x_k| < a, |x_{k-1}| < a, \ldots, |x_1| < a\} \qquad (2)$$

If the shocks are independent with the same cumulative distribution function $F_x(x)$, Eq. (2) becomes (with $\overline{H}_T(0) = 1$):

$$\overline{H}_T(k\Delta t) = [F_x(-a,a)]^k = [F_x(a) - F_x(-a)]^k \qquad (3)$$

2.2 Additive drift

Let us introduce the random variable s_k representing the cumulative drift up to the k-th shock:

$$z(k\Delta t) = s_k = \sum_{i=1}^{k} x_i \qquad (4)$$

With the above notation:

$$\overline{H}_T(k\Delta t) = \Pr\{|s_k| < a, |s_{k-1}| < a, \ldots, |s_1| < a\}\,\} \qquad (5)$$

If the amplitude of the shocks is constrained to be a positive random variable (for example wearing), the survival function (5) can be simplified [4]. In the more general case of positive as well as negative shocks, the multivariate distribution function of $(s_1, \ldots s_k)$ has to be considered and (5) has to be evaluated by means of a k-dimensional integral limited between *-a* and *+a*.

2.3 Random arrival times

Let's now suppose that the shocks occur randomly in time according to a point process *N(t)*. The process is completely specified if the probability $P_k(t)$ of having k shocks at time t is known:

$$P_k(t) = \Pr\{N(t) = k\} \qquad (6)$$

The survival probability becomes:

$$\overline{H}_T(t) = \sum_{k=0}^{\infty} \overline{H}_T(t/k) \cdot \Pr\{N(t) = k\} \tag{7}$$

where $\overline{H}_T(t/k)$ is the conditional probability of surviving k shocks at the time t. $P_k(t)$ could be, for example, the probability of k shocks in a Poisson process. In the following the analysis is restricted to equispaced shocks. The case of random arrival times was examined in [4].

3. Computational example

The sequence of shocks can be interpreted as the fluctuations in the value of the observable parameter versus time. If the fluctuations result from the combined effect of different independent disturbances, their amplitude distribution turns out to be Gaussian. Hence, we performed some numerical computations of the formulas derived in Section 2, by adopting a Gaussian distribution for the amplitude x with expected value μ=0 and standard deviation σ whose density has the following expression:

$$f_x(x) = \frac{1}{\sqrt{2\pi}\sigma} e^{-\frac{x^2}{2\sigma^2}} \tag{8}$$

In Figure 1 and 2, we have reported the survival function $\overline{H}_T(t)$ for the non-cumulative (Eq. 3) and the additive drift (Eq. 5), respectively, for different values of the tolerance threshold a (normalized to the standard deviation σ). From the curves of Fig. 1 and 2, we can derive the calibration interval θ. Given the acceptance level α, we obtain θ as the intercept of the survival curves at a level equal to α. The results are in Fig. 3 for two acceptance levels α=0.95 and α=0.99 for both the non-cumulative and the additive model. Plots of Fig. 3 provide a mean for designing the calibration procedure of the instrument under consideration.
For instance, we derive from Fig. 3 that for a=3σ and with an acceptance level α=0.99, we must assume a calibration interval θ =3 days in the case of non-cumulative drift or θ =1 day in the case of additive drift.

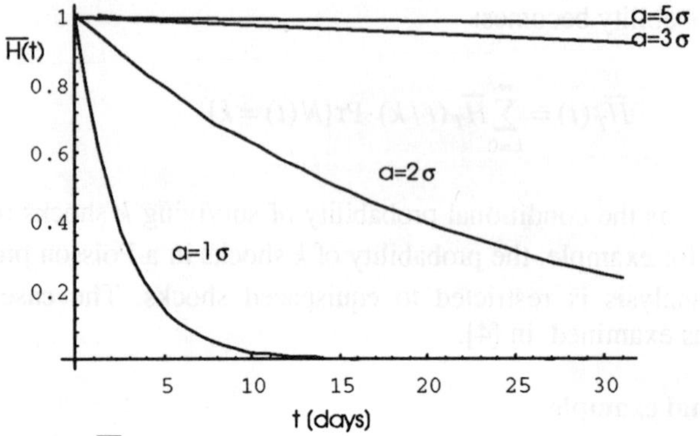

Fig.1 Survival function $\overline{H_T}(t)$ versus time for different values of *a*. Non-cumulative Gaussian model

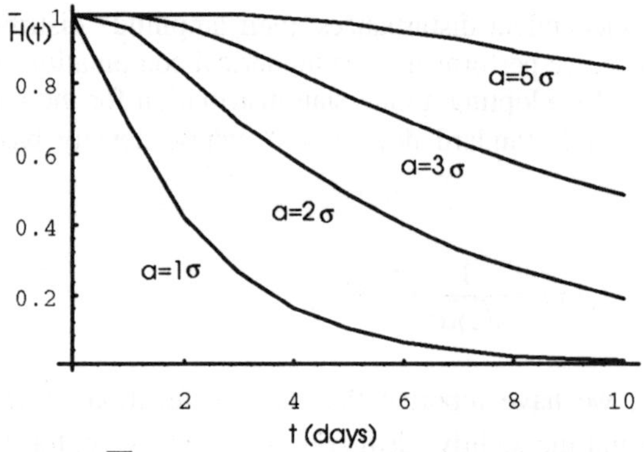

Fig. 2: Survival function $\overline{H_T}(t)$ versus time for different values of *a*. Additive Gaussian model

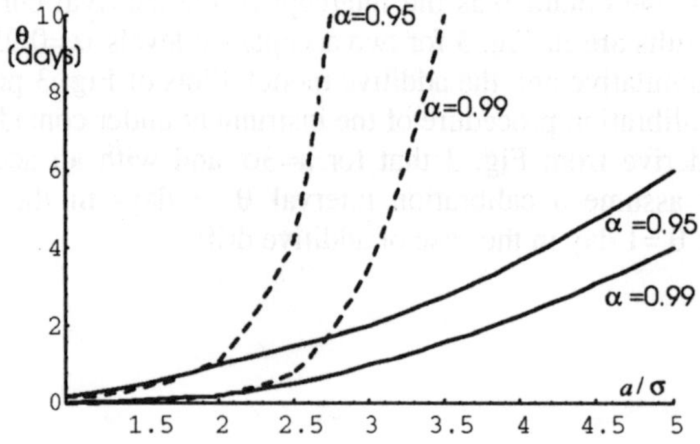

Fig.3 Calibration interval θ versus the threshold *a* for two acceptance levels α (dashed line = non-cumulative case, solid line = additive case)

4. Experimental Results

In order to have a preliminary validation of the theoretical models developed in the previous sections, we collected a sample of experimental data obtained from the calibration tests performed on three similar instruments. The instruments are employed to measure wafer thickness (of the order of hundreds of μm). A calibration test is performed at the beginning of every work shift (three times a day). The test consists in performing a measurement on a reference wafer of known thickness and reporting the deviation between the reference and the measured thickness values (in μm) on a control chart. When the operator finds the instrument out-of-the tolerated band, a suitable adjustment procedure is initiated. Since the considered parameter is a deviation, the correct value should be 0, and the calibration band is assumed symmetric and bilateral. Consistently with the measured parameter, the value of the threshold a is set in μm. Due to the homogeneity of the three instruments and of the measuring conditions, the calibration data arising from the three instruments have been merged into a single accumulated sample representative of that class of instruments. After collecting data for a period of two months, the accumulated experimental sample consisted in N=392 measures.

The experimental sample has been submitted to a spectral analysis, by examining the autocorrelation function. The result of this analysis showed that the measured deviations have a white spectrum indicating that there is no correlation between successive measures, therefore the non-cumulative drift model (Sec. 2.1) seems the more appropriate. The sample of the collected deviations was fitted by a Gaussian density providing the following results: mean value $\mu = -0.55$ μm and standard deviation $\sigma = 1.1$ μm. Due to the size of the sample, the relative uncertainty of the estimated mean value is 10%. From the control charts, we evaluated the time intervals between two successive out-of-tolerance conditions (measured in number of work shifts) for the three instruments under examination. The empirical survival function of this data sample is reported in Fig. 4 (dotted line) together with the sf of the non-cumulative model (3) using the best fit Gaussian distribution as the cdf $F_x(x)$ of each single shock (solid line). The hypothesis that the experimental sample belongs to the family expressed by the theoretical model (3), with $F_x(x)$ given by the fitted Gaussian distribution, was submitted to a Kolmogorov Smirnov goodness-of-fit test. The hypothesis cannot be rejected at a significance level of 0.9. Hence, we conclude that the agreement is satisfactory. The optimal calibration interval θ (in workshift units) as a function of the tolerated band a and the acceptance level α was then estimated according to (1) and is reported in [4].

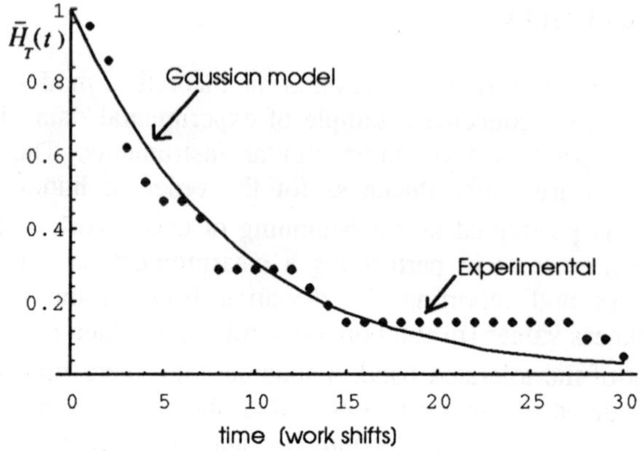

Fig. 4 The empirical survival function $\overline{H}_T(t)$ and the theoretical one as evaluated from (3)

5. Conclusions and Future Work

The paper presented various stochastic drift models that can be adapted to model the out-of-tolerance condition of a measuring instrument. From these models, the distribution of the time to reach an out-of-tolerance condition can be evaluated. The optimal calibration interval can then be inferred such that the risk of performing incorrect measurements is less than a preassigned level. A preliminary validation on experimental data is presented. The directions for future work are twofold. First, we intend to apply the model to more extensive classes of instruments, and to this aim we need to collect statistically significant samples of experimental data. The second direction consists in refining the model, for instance resorting to the idea of a combination of stochastic and deterministic drift and in adding cost functions [5] quantifying the cost of a wrong measure and the cost of the process unavailability.

References

[1] OIML Organisation Internationale de Métrologie Légale - DI 10, 1984.

[2] NCSL Tech. Rep. National Conference of Standard Laboratories RP1, 1989.

[3] J.D. Esary, A.W. Marshall, F. Proschan *The Annals of Probability*, 1, 627-649, 1973.

[4] A. Bobbio, A. Montefusco, P. Tavella, Proc. IEEE-IM Tech. Conf., Bruxelles, 54-57, June 1996.

[5] K. Dirscherl, L. Nielsen, B. Ostergaard, E. Bak-Sorensen, K. Schüsler, Danish Inst. Fundamental Metrology, DFM-96-R19 Tech. Rep. 1996.

Advanced Mathematical Tools in Metrology III
Edited by P. Ciarlini, M.G. Cox, F. Pavese & D. Richter

PROJECTION METHOD FOR LIDAR MEASUREMENT

C. BÖCKMANN
Universität Potsdam, Institut für Mathematik, Am Neuen Palais 10, D-14415 Potsdam, Germany

The linear inverse ill-posed problem of aerosol size distribution determination from extinction measurements was solved successfully with a projection method.

1 Introduction

Atmospheric aerosols play an important role in many atmospheric processes, e.g. in the complex dynamics of tropospheric ozone production and of stratospheric ozone destruction because on the surface of such aerosols, there are chemical reactions which activate chlor from CKW and FCKW. One of the key aspects in a further understanding of the importance of aerosols is the investigation of the spatial and temporal variability of microphysical properties of the particles, e.g. parameters describing their mean size, their mass and surface-area concentrations. The aim of multiple-wavelength LIDAR-measurements is for instance the determination of the aerosol size distributions. We consider the problem of determining a suitable distribution $n(r)$ from a set of N measurements $\alpha^{Aer} = (\alpha(\lambda_1), \ldots, \alpha(\lambda_N))^T$. Assuming that the aerosol polydispersion consists of Mie particles of varying radius r and a fixed refractive index m, $m = 1.5$, then

$$\alpha^{Aer}(\lambda_j) = \int_{r_1}^{r_2} \pi r^2 Q_{ext}(\lambda_j, r; m) n(r) dr = \int_{r_1}^{r_2} k(\lambda_j, r) n(r) dr = (Kn)_j \quad (1)$$

with $j = 1, \ldots, N$, where Q_{ext} is the Mie theory extinction efficiency factor, $n(r)$ is the unknown aerosol size distribution function we wish to determine and $\alpha^{Aer}(\lambda)$ is the data function. The projection method proposed here does not require a trial distribution or an initial guess to achieve results.

2 The Projection Method

The inverse operator K^{-1} is not continuous, i.e. the problem is ill-posed and we need a suitable regularization method. Typically, these problems are solved by projection methods. We follow the idea of Luois[2]. We select a function system $X_M = \text{span}\{\varphi_1, \ldots, \varphi_M\}$ with $\varphi_i(r) = \begin{cases} 1 & r \in [(i-1)h, ih] \\ 0 & \text{else} \end{cases}$, $h = 1/M$ and $i = 1, \ldots, M$. Then we change the mathematical model with projectors into

$Q_M K P_M n^N = Q_M \alpha$, i.e. we get a linear (underestimated) equation system $Bn^N = b(\alpha^N)$. The solution is $n(r) = \sum_{i=1}^{M} n_i^N \varphi_i(r)$. We carried out a numerical simulation using the following procedure. First, we assumed a distribution $n(r) = \exp(-0.5 \ln^2(r/r_{med})/\ln^2(\sigma))$ with $r_{med} = 0.18$ and $\sigma = 1.75$, see Figure 1 (a)-(c) (dotted lines), and second, we calculated the extinction coefficients α^{Aer} for 30 (10, 6) equidistant (not a condition) wavelengths between 300 and 900 nm. The distribution thus reconstructed (solid lines) was compared with the given one (dotted lines). As regularization tool for the linear system we used truncated singular value decomposition with $\delta = 0.0001(0.03, 0.03)$. We require 3 mantissa digits of α_i^N, i.e. only small measurement errors are possible. Also compare the later papers Böckmann [1] and Müller [3] there are even better results, e.g. by application of Mollifier-methods.

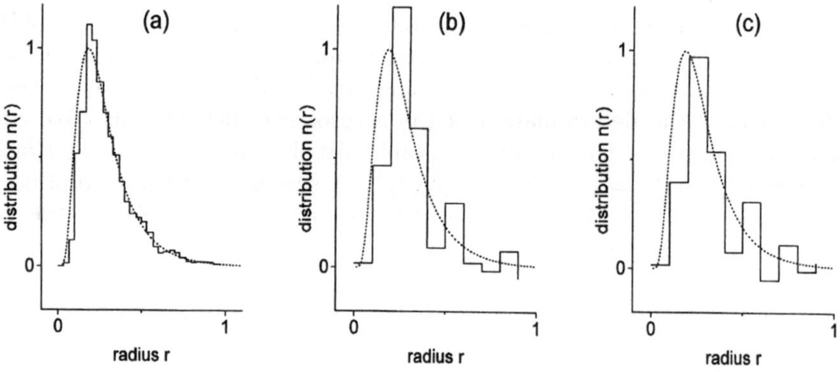

Figure 1: Results for 30(10,6) noisy data with 2(2,3) mantissa digits (a)((b),(c)) solid lines.

References

1. C. Böckmann and J. Niebsch in *Advances in Atmospheric Remote Sensing with Lidar*, ed. A. Ansmann *et al*, (Springer, Berlin, 1996).
2. A.K. Louis, *Inverse und schlecht gestellte Probleme*, (BG Teubner, Stuttgart, 1989).
3. D. Müller *et al* in *Advances in Atmospheric Remote Sensing with Lidar*, ed. A. Ansmann *et al*, (Springer, Berlin, 1996).
4. R. Neuber *et al*, *Ber. Bunsenges. Phys. Chem.* **96**, 351 (1992).

Advanced Mathematical Tools in Metrology III
Edited by P. Ciarlini, M.G. Cox, F. Pavese & D. Richter

PHOTON FLUX MEASUREMENTS BY REGULARISED SOLUTION OF INTEGRAL EQUATIONS

S. CASTELLETTO
Dipartimento di Ingegneria Elettronica - Politecnico di Torino
c.so Duca degli Abruzzi, 24 - I 10100 Torino (Italy)
M.L. RASTELLO
Istituto Elettrotecnico Nazionale G. Ferraris - Torino (Italy)
Strada delle Cacce, 91 - I 10135 Torino (Italy)
E-mail : Rastello@piccolo.cstv.to.cnr.it

The spectral photon flux of a parametric fluorescence beam is evaluated through a system of integral equations. The technique permits to overcome the drawbacks intrinsic to the detection of very low signals. The experimental results show the capabilities of the method with respect to standard techniques.

1 Introduction

Photon fluxes can be measured in principle by an absolute detector whose spectral responsivity is $W(\lambda) = \frac{1}{h\nu} = \frac{\lambda}{hc}$. Actually, this measurement can be performed by a quantum-flat trap detector[1], whose absolute spectral responsivity $S(\lambda) \approx W(\lambda)$ apart from a correction factor taking into account some losses in the detector. As to the beam, parametric fluorescence (PF) in a non-linear crystal is a suitable photon source as its spectral emmission just lies in the wavelength range where trap detectors are quantum-flat. Its theoretical spectral distribution[2] is given by:

$$\Phi_{th,\lambda} = \frac{2(2\pi)^3}{\epsilon_o} \frac{P_p}{{n_p}^2{\lambda_p}^2} L{d_{eff}}^2 \frac{(\lambda - \lambda_p)^2}{\lambda^6} \tag{1}$$

where the pump is a laser with wavelength λ_p and power P_p, the non linear crystal has length L, refraction index n_p and effective non-linear coefficient d_{eff}, and ϵ_o is the vacuum dielectric constant.

The measurement of PF spectral flux is particularly difficult because the emission of generated photons is in a cone of known aperture around pump direction and each wavelength is spectrally separated from the other. Moreover very low signals have to be measured in the presence of a significant straylight and standard techniques involving monochromators can hardly be applied since PF power is 10^6 times lower than P_p. To overcome these intrinsic drawbacks, a dedicated experimental system has been set up and the so-called broadband filter (BF) method[3] has been used. The main advantages of the BF method in this case are that the emission cone is considered altogether and filter signals are considerably higher than monochromator ones. The spectral photon flux $\Phi_p(\lambda)$ of PF cone has been evaluated by measuring N output signals w_n^s ($n = 1, \ldots, N$) coming from the trap detector when it is irradiated by the source combined in turn with N broadband filters with known spectral transmittances $\tau_n(\lambda)$ ($n = 1, \ldots, N$). Then, measurement data are analyzed by solving the system of N integral equations:

$$w_n^s = \int_{\lambda_a}^{\lambda_b} \Phi_p(\lambda)\tau_n(\lambda)\mathrm{d}\lambda \quad \text{for} \quad n = 1, 2, \ldots, N \tag{2}$$

in order to extract the required spectral information on $\Phi_p(\lambda)$. From a mathematical point of view, a broad-band integral observation involves the unknown function through a known function, the kernel, representing the physical action of the measurement system, the filter in this case. The pass-band of the filter is generally rather broad. Unfortunately, the smoothness of $\tau_n(\lambda)$ causes the ill-conditioning of system (2) and data w are affected by experimental data errors. Thus, a regularized formulation can be applied to the BF method, and a significant solution of system (2) obtained by adding theoretical informations on the unknown curve $\Phi_p(\lambda)$. The regularized formulation allows the spectral distribution to be reconstructed quite easily as shown by the reported experimental results.

2 Proposed solution to the problem

For obtaining a significant solution of the system (1), a regularization method[4] is introduced to reduce the set of infinitely many approximated solutions of system (2), to those having some useful properties in common with the expected solution to the problem. This method works well as theoretical expression (1) allows us to construct a suitable model $\Phi_{th}(\lambda)$ of the spectral fluorescence $\Phi_p(\lambda)$. As $\Phi_p(\lambda)$ and $\Phi_{th}(\lambda)$ are characterized by the same amount of smoothness, they are represented by different linear combinations of the same basis functions. In particular, $\Phi_p(\lambda)$ is described by a spline function $X(\lambda)$. Interval $[\lambda_a, \lambda_b]$ is partitioned into subintervals $[\hat{\lambda}_m, \hat{\lambda}_{m+1}]$ ($m = 1, 2, \ldots, M-1$) by a mesh of M knots $\hat{\Lambda} \equiv \{\hat{\lambda}_1, \hat{\lambda}_2, \ldots, \hat{\lambda}_M\}$, with $\hat{\lambda}_1 = \lambda_a$ and $\hat{\lambda}_M = \lambda_b$. In each subinterval, spline[5] $X(\lambda)$ is defined as a polynomial of degree not exceeding 3; polynomials related to adjacent subintervals are linked together by making $X(\lambda)$ continuous with its first two derivatives $\mathrm{d}X(\lambda)/\mathrm{d}\lambda$ and $\mathrm{d}^2X(\lambda)/\mathrm{d}\lambda^2$ at the internal knots $\hat{\lambda}_2, \ldots, \hat{\lambda}_{M-1}$. Moreover $\mathrm{d}^2X(\lambda)/\mathrm{d}\lambda^2$ is assumed to be zero at the extreme knots $\hat{\lambda}_1$ and $\hat{\lambda}_M$. Consequently, $X(\lambda)$ has M degrees of freedom: they are represented by the values of its samples $X(\hat{\lambda}_m)$

$(m = 1, 2, \ldots, M)$ at the knots. To emphasize this univocal dependence, the set of cardinal splines $b_m(\lambda)$ $(m = 1, 2, \ldots, M)$ is defined as $b_m(\hat{\lambda}_k) = \delta_{km}$, $(k = 1, 2, \ldots, M,\ m = 1, 2, \ldots, M$ where δ_{km} is the Kronecker delta. The M samples of each $b_m(\lambda)$ are null except for the m-th sample equal to 1. Cardinal splines are grouped in the M-vector $\boldsymbol{b}(\lambda) \equiv [b_1(\lambda)\, b_2(\lambda) \ldots b_M(\lambda)]^{\mathrm{T}}$. Thus, curve $\Phi_p(\lambda)$ is conveniently represented with a negligible approximation error by the spline $X(\lambda) = \boldsymbol{x}^{\mathrm{T}}\, \boldsymbol{b}(\lambda) = \sum_{m=1}^{M} x_m\, b_m(\lambda)$ where vector $\boldsymbol{x}^{\mathrm{T}} \equiv [x_1\, x_2 \ldots x_M]$ groups the unknown coefficients, characterizing the unknown $X(\lambda)$ as a linear combination of the M known cardinal splines.

Analogously, the model is represented by combining the same cardinal splines as $\Phi_{th}(\lambda) = \xi[\hat{\boldsymbol{x}}]^{\mathrm{T}}\boldsymbol{b}(\lambda)$, where $\hat{\boldsymbol{x}}$ groups the known coefficients describing the model, and scaling factor ξ takes into account the fact that, generally, the model is known apart from an unknown factor.

By replacing $\Phi_p(\lambda)$ with $\Phi(\lambda)$ in the integrals appearing on the right-hand side of eq. (2) the algebraic system of equations can be rewritten with vectorial notation:

$$\boldsymbol{w}^x = \boldsymbol{K}\,\boldsymbol{x} \tag{3}$$

where $\boldsymbol{w}^x \equiv [w_1^x\, w_2^x \ldots w_N^x]^{\mathrm{T}}$ and $\boldsymbol{K}$ is an $N \times M$ matrix whose elements are

$$k_{nm} = \int_{\lambda_a}^{\lambda_b} \tau_n(\lambda) b_m(\lambda) \mathrm{d}\lambda \quad n = 1, 2, \ldots, N;\ m = 1, 2, \ldots, M. \tag{4}$$

The ill-conditioning of the system (3) strictly depends on matrix $\boldsymbol{K}$. It is enhanced when the bandwidths of the filters are large compared to the inter-knot interval, i.e. for increasing M. On the other hand, too low a number, M, of knots may introduce an unacceptable error in approximating $\Phi_p(\lambda)$ by $\Phi(\lambda)$. The unknown M-vector $\boldsymbol{x}$ has to be determined in such a way that $\boldsymbol{w}^x$ reproduces the measured N-vector $\boldsymbol{w}^s \equiv [w_1^s\, w_2^s \ldots w_N^s]^{\mathrm{T}}$. These are affected by statistically independent random errors with mean value zero and standard deviation σ_n, whereas the elements of matrix $\boldsymbol{K}$ are assumed to be exactly known. As in the ordinary weighted least squares problems, the searched solution $\boldsymbol{x}$ is asked to minimise the euclidean norm of the errors normalized by variances $\mathcal{Q}^2(\boldsymbol{x}) = [\boldsymbol{K}\boldsymbol{x} - \boldsymbol{w}^s]^{\mathrm{T}}\boldsymbol{U}^{-1}[\boldsymbol{K}\boldsymbol{x} - \boldsymbol{w}^s]$, where $\boldsymbol{U} \equiv \mathrm{diag}\,[\sigma_1^2, \sigma_2^2, \ldots, \sigma_N^2]$ is the diagonal matrix of variances.

Moreover the system is underdetermined $(N < M)$ because a low number of filters has been used in the experiment; consequently, the squared weighted Euclidean norm $\mathcal{G}^2(\boldsymbol{x}, \xi) = [\boldsymbol{x} - \xi\hat{\boldsymbol{x}}]^{\mathrm{T}}[\boldsymbol{Z}]^{-1}[\boldsymbol{x} - \xi\hat{\boldsymbol{x}}]$ is introduced as figure of similarity between the unknown curve $\Phi(\lambda)$ and the model $\Phi_{th}(\lambda)$; matrix $\boldsymbol{Z}$ takes into account the uncertainty within which the model is known. The global squared norm $\mathcal{F}^2$ to be minimized is given by $\mathcal{F}^2(\boldsymbol{x}, \xi, \alpha) = \mathcal{Q}^2(\boldsymbol{x}) +$

$\alpha\, \mathcal{G}^2(\boldsymbol{x},\xi)$, where α is the (positive) regularization parameter, trading-off the regularization action of norm $\mathcal{G}(\boldsymbol{x},\xi)$ (large α) and the fidelity to measured data proper of norm $\mathcal{Q}(\boldsymbol{x})$ (small α). First, the new unknown $(M+1)$-vector $\boldsymbol{y} \equiv [\boldsymbol{x}^{\mathrm{T}}\,\xi]^{\mathrm{T}}$ is introduced by augmenting vector $\boldsymbol{x}$ by scalar ξ. As shown in Ref. 3, for a given α, the analytical solution minimizing norm $\mathcal{F}(\boldsymbol{x},\xi,\alpha)$ is given by

$$[\tilde{\boldsymbol{x}}(\alpha)^{\mathrm{T}}\,\tilde{\xi}(\alpha)]^{\mathrm{T}} = \begin{bmatrix} \boldsymbol{K}^{\mathrm{T}}\boldsymbol{U}^{-1}\boldsymbol{K} + \alpha\boldsymbol{Z}^{-1} & -\alpha\boldsymbol{Z}^{-1}\hat{\boldsymbol{x}} \\ -\alpha\hat{\boldsymbol{x}}^{\mathrm{T}}\boldsymbol{Z}^{-1} & \alpha\hat{\boldsymbol{x}}^{\mathrm{T}}\boldsymbol{Z}^{-1}\hat{\boldsymbol{x}} \end{bmatrix}^{-1} \times \\ \times \begin{bmatrix} \boldsymbol{K}^{\mathrm{T}}\boldsymbol{U}^{-1}\boldsymbol{w}^{s} \\ 0 \end{bmatrix} \qquad (5)$$

and the optimal value of α corresponds to the best attainable accuracy in the numerical inversion in eq. (5).

3 Experimental results

PF has been generated by pumping a cube of 10 mm length lithium iodate with an Argon laser. Spectral photon fluxes have been measured by a trap detector equipped, in turn, with $N = 7$ commercial interference filters every 50 nm from 450 nm to 750 nm. Integrals in eqs. (4) have been evaluated with sufficient accuracy by interpolating with a cubic spline the measured values of $\tau(\lambda)$, sampled in a mesh of knots much denser than $\hat{\Lambda}$. In this way, errors are negligible with respect to the errors intrinsic to the BF method. The relative uncertainties σ_n/w_n^s $(n = 1, 2, \ldots, N)$ of the measured data are varying in the range 0.5 to 1%. The model $\Phi_{th}(\lambda)$ is the nominal curve (1) and $M = 11$. At the pumping wavelength $\lambda_p = 351$ nm, the spectral flux has been measured by means of two completely different measurement systems. First, a direct method has been used with a spectrometer. Then, the BF method has been implemented and the obtained results are displayed in Fig 1, where squares refer to the spectral flux measured by the spectrometer, solid line is model $\Phi_{th}(\lambda_p = 351$ nm), and circles refer to the reconstructed BF solution. This result shows clearly that BF method allows us to obtain much more reliable informations on the actual behaviour of very low spectral fluxes, as in PF.

4 Conclusion

A regularized solution of the BF method is presented for spectral photon flux measurements. The formulation involves the theoretical knowledge of the curve to be measured and allows one to reconstruct quite easily the spectral function.

Freely offered computer facilities by the CSTV of the CNR are acknowledged.

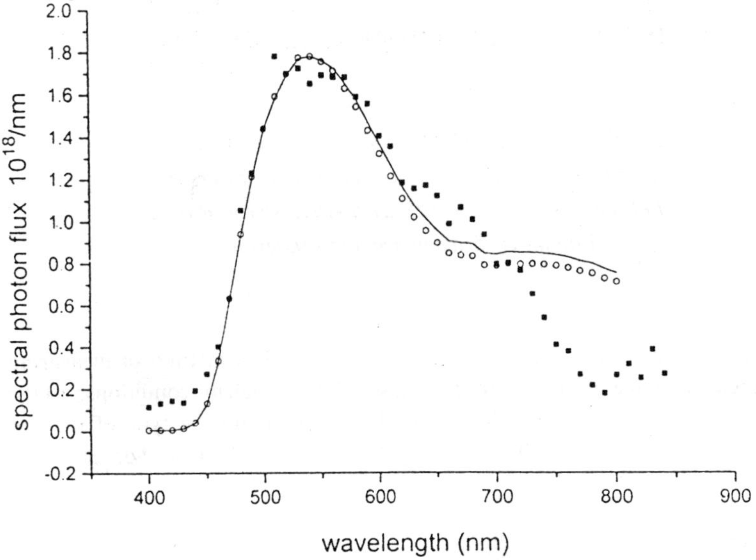

Fig. 1 Diagram of $\Phi_p(\lambda)$ (squares), $\Phi_{th}(\lambda)$ (solid line), and the reconstructed $\Phi(\lambda)$ (circles) versus λ for $\lambda_p = 351$ nm.

5 References

1. N. P. Fox, Metrologia **28**, 197 (1991).

2. K. Koch, E.C. Cheung, G.T. Moore, S.H. Chakmakjian, J.M. Liu, IEEE Journ. QE , **31**, 769 (1995).

3. M.L. Rastello, A. Premoli, and G. Bonicatto, IEEE Trans. IM **45**, 22 (1996).

4. S. Denasi, A. Premoli and M.L. Rastello in *Advanced Mathematical Tools in Metrology* , eds. P. Ciarlini, M.G. Cox, R. Monaco, and F.Pavese (World Scientific, Singapore, 1994).

5. J.H. Ahlberg, E.N. Nilson and J.L. Walsh in *The Theory of Splines and Their Applications* (Academic Press, New York, 1967).

Advanced Mathematical Tools in Metrology III
Edited by P. Ciarlini, M.G. Cox, F. Pavese & D. Richter

CORRECT SOLUTIONS OF FIT PROBLEMS IN DIFFERENT EXPERIMENTAL SITUATIONS

Y.V. CHEBRAKOV, V.V. SCHMAGIN
Department of Mathematics, Technical University,
Politehnitsheskaja 29, 195221 St-Petersburg, Russia
E-mail: chebra@phdeg.hop.stu.neva.ru

In this paper we prove a) LSM-solution is incorrect if a method of measuring dependent variable is added to the classical fit problem conditions; b) the procedure of correct solving the fit problem depends on the type of realised experimental situation identified by imaginary device method (see Sec. 2).

1 Introduction

The main goal of this investigation is to analyse the dependence existing between a certain characteristic y of a given object and a set of characteristics x; in other words, it is required first to determine the experimental dependence $y = y(x)$ and then to fit it by an analytic curve $F(\mathbf{B}, x)$, where $\mathbf{B}$ is a vector parameter of the function $F(\mathbf{B}, x)$. In regression analysis theory[1] it is proved that

1) if for the data array $\{y_n, x_n\}$ the conditions in the Gauss-Markov theorem are satisfied, then the least square method (LSM) estimate is the best among all (linear and non-linear) unbiased estimates;

2) as a rule, it is impossible in real experimental conditions to test that all the requirements of the Gauss-Markov theorem are satisfied (that is why in regression analysis there is a set of non-traditional estimate methods).

In this paper (see Sec. 2 and 3) we present a set of methods by which one may

a) test whether the conditions in the Gauss-Markov theorem are satisfied (identify the realised operational situation of a given experiment);

b) solve fitting problems with guaranteed interpolative accuracy;

c) search the best fitting model among a given set of models.

2 Identification of the realised operational situation

As stated[2, 3] the strategy of data processing by regression methods depends on a concrete type of realised experimental situation. Namely, for any fitting function $F(x)$ it is necessary to distinguish three cases:

1) the data array is obtained by the "round-off" procedure for $F(x)$, that is, for this function the accuracy of the computed values y_n corresponds to the accuracy of the values y_n in the initial data array for all n;

2) the function $F(x)$ with a mixture of $\{e_n\}$, where $\{e_n\}$ is a noise or "a set of random variables"; however, by deleting several elements from the initial data array, we may obtain a truncated data array that satisfies the conditions in 1 above (*removable* or *local inadequacy* of the function is realised);

3) the same as in 2, but either the number of deleted elements from the initial data array is too large (in the investigator's opinion) or one cannot obtain a truncated data array (*irremovable* or *global inadequacy* of the function).

Since in a real (experimental) situation all values of y_n are measured with finite accuracy, we introduce an *imaginary* device by which, if necessary, we may "remeasure" any values of y_n. We denote by α the value of the minimal point of its scale. Using the imaginary device for a fixed n_0, we determine the least value α_{min} for which on a set of plans $\{\mathbf{X}\}$, containing all possible combinations of N, $N-1$, ... , $N-n_0$ observations, the inequality $|g(y_n) - g(F(\mathbf{B}', x_n))| \leq \alpha_{min}$, holds, where $\mathbf{B}'$ is the LSM estimate of the vector parameter and $g(y)$ is a *measuring* function of the imaginary device, N is the general number of observations. It is evident that for any fitting function $F(x)$, using the above method and a set of combinatorial procedures that allow us to generate a set of plans $\{\mathbf{X}\}$, we may identify the type of realised situation by the value of α_{min}.

3 Finding the best even-approximating multinomial

1. *Operational situation number 1.* Suppose that the values of a continuous and sufficiently smooth function $y = F(x)$ are given within ε on a discrete plan $\{x_n\}$. Then it can be proved that a) the solution of classical regression analysis is incorrect; b) the correct (according to Hadamard) solution of the above problem may be obtained by means of approximation theory or the imaginary device method.

a) We adduce a simple explanation why the classical solution is not the best in the given case. Indeed, in the imaginary device method for approaching operational situations of real experiments the *classical* model of data analysis $y_n=F(x_n)+e_n$ is replaced by a new model $y_n=g(F(x_n)+e_n)$. It is evident that one may return to the classical model if not only random noise but a systematic error (model error) are included in $\{e_n\}$ in the structure of this model. But in this case the Gauss-Markov theorem requirements are not satisfied.

b) As is well known in approximation theory[4], if the type of function $F(x)$ is given and either the $(m+1)$th derivative of this function varies weakly on the plan $\{x_n\}$, or the function $F(x)$ has the form of an even-converging power series on the plan $\{x_n\}$ in $[-1, 1]$, then the problem of finding the best even-approximating multinomial for the discrete dependence $\{y_n, x_n\}$ offers no difficulty. Indeed, in the first case the solution of the problem is an interpolative multinomial $P_m(x)$ with a set of Chebyshev points (this multinomial is close to the best even-approximating one). In the second case, one may obtain the solution by an economical procedure of an even-converging power series. If the above conditions for $F(x)$ are not satisfied or, for example, a function of this type is unknown, then one may obtain the correct solution of the given problem by means of the imaginary device method (see Sec. 2). In this case the solution has the form $g(P_m(x))$.

2. *Operational situation number 2.* In this situation[2,3] if one wants to find the best fitting multinomial, then it is necessary to delete all outliers in the initial data array.

3. *Operational situation number 3.* As is generally known[5], the estimated values of the inadequacy parameters of the fitting model might be strongly affected by the form of the postulated $(F(\mathbf{B}, x))$ and true $(G(\mathbf{C}, x))$ fitting models, and by the plan of the experiment that is determined by given magnitudes of the vector x. It is a traditional standpoint that the dependence on the estimated values of $\mathbf{B}$ in the above operational conditions of the experiment makes the solution of the problem of restoring the parameters on the given array $\{y_n, x_n\}$ considerably more difficult. We prove that the above dependence helps a) to establish the inadequacy of the model $F(\mathbf{B}, x)$; b) to restore the values of the parameters $\mathbf{B}$ in spite of the inadequacy of this model.

To prove we assume that the form of the functions $F(\mathbf{B}, x)$ and $G(\mathbf{C}, x)$ is known and that, for any plans of an experiment, a_1 and a_2 are, respectively, the minimal and maximal values of the independent variable x. Then, for any plans of the experiment, we may first determine, by the method of least squares, with the required exactness, the magnitude of *the discrete theoretical ratio* existing between the parameters of the functions $F(\mathbf{B}, x)$ and $G(\mathbf{C}, x)$; secondly, using the estimated values of the parameters **B** and **C**, we can calculate *the discrete experimental magnitude* of the same ratio; thirdly, we can find the magnitude of the same ratio at infinity, that is, in the case when both functions are given on the continuum of the segment $[a_1, a_2]$ (*the continuous theoretical ratio*). It is evident that by comparing the three ratio magnitudes we can determine the adequacy of the function $G(\mathbf{C}, x)$ and the inadequacy of the function $F(\mathbf{B}, x)$. The exact solution of the problem of restoring the parameters of the function $F(\mathbf{B}, x)$ can be obtained from the discrete theoretical ratio by means of integral or numerical methods. It should be noted that, in practice, the form of $G(\mathbf{C}, x)$ may be unknown. In this case we may select the best fitting multinomial as $G(\mathbf{C}, x)$.

Example. Let $F(\mathbf{B}, x)$ be a linear model, $G(\mathbf{C}, x)$ a model that can be generated by characteristic numbers λ_m from a four degree multinomial $P_4(\mathbf{C}, x)$. Then for the linear model b_0+b_1x instead of the LSM estimates of its parameters we use estimates computed by means of the formulae

$$b_0 = c_0' - \lambda_2 c_2'(a_c^2 + 2a_u)/6 - \lambda_3 c_3' a_c(a_c^2 + a_u) - \lambda_4 c_4' (a_c^4 - a_u^2)/5, \qquad (1)$$
$$b_1 = \lambda_1 c_1' + \lambda_2 c_2' a_c + \lambda_3 c_3' (9a_c^2 - 6a_u) + \lambda_4 c_4' (4a_c^3 - 6 a_c a_u),$$

which derived from the LSM integral equations, where $a_c = a_1 + a_2$, $a_u = a_1 a_2$, c' – LSM estimates of multinomial coefficients. If function $2x^{1,5}$ is determined on plan $\{x_n\} = (2, 3, \ldots, 10)$ then for the linear model b_0+b_1x a) $\mathbf{B}_{LSM} = (7{,}239;\ -11{,}954)$; b) $G(\mathbf{C}, x)=P_4(\mathbf{C}, x)$ and $\mathbf{B}_{C\ LSM} = (7{,}2607{:}-12{,}507)$, where $\mathbf{B}_{C\ LSM}$ is calculated by Eq. 1; c) $\mathbf{B}_{S\ LSM} = (7{,}2612{:}-12{,}510)$, where $\mathbf{B}_{S\ LSM}$ is calculated from the LSM integral equations; in other words, when function $2x^{1,5}$ is determined on the continuum of the segment [2, 10].

References

1. G.A.F. Seber, *Linear regression analysis* (J. Wiley, N. Y., 1977).
2. Y.V. Chebrakov, *System analysis methods in experimental inquiries* (St.Petersburg State Univ.Press, St.Petersburg, 1997).
3. Y.V. Chebrakov, V.V. Shmagin, *Izv. Vyzov. Ser. Fiz.* **5**, 40 (1995).
4. P.J. Laurent, *Approximation et optimisation* (Hermann, Paris, 1972).
5. K.R. Draper, H. Smith, Applied regression analysis (J. Wiley, N. Y., 1981).

Advanced Mathematical Tools in Metrology III
Edited by P. Ciarlini, M.G. Cox, F. Pavese & D. Richter

An algorithm for the nonlinear TLS problem in polynomial fitting

PATRIZIA CIARLINI
Istituto per le Applicazioni del Calcolo "M. Picone", CNR
Rome, Italy

MARIA LAURA LO CASCIO
Dip. Me.Mo.Mat. Università "La Sapienza"
Rome, Italy

To fit a curve to data affected by errors in both variables is a common problem in metrological applications. This problem has been completely solved from a theoretical and computational point of view in the linear case. In this paper a polynomial of degree n is chosen to model the data behaviour and an iterative algorithm, particularly suitable to metrological data, is described, based on polynomial root finding. Its application to metrological data is finally discussed.

1 Introduction

Several applications involve fitting of datapoints (t_i, y_i), i=1,2,...,m to a curve; the most commonly used solution is the one obtained by means of the Least Squares estimator, where it is assumed that the values t_i are error free, while the observations y_i only are affected by errors. Unfortunately in many metrological situations these assumptions are not realistic being the values of the indipendent variable t as well as the observations obtained by measurements.

To overcome this problem the Total Least Squares approach has been considered. It was firstly introduced by Golub and Van Loan[1] and subsequentely studied to construct lines or hyperplanes from theoretical point of view[2,3]. Algorithms were built based on the singular value decomposition[3] or on a geometrical formulation of the problem[4]. It was also shown that the TLS approach gives better estimates of the model parameters than the LS method, under the hypotesis that the errors are uncorrelated random variables with zero mean and equal variance. In the TLS formulation the existence and/or uniqueness of a solution are not always assured. A nonlinear TLS problem arises from the fitting of observations to a nonlinear model $\Phi(\bullet,\alpha)$, where α. is the vector of the parameters. In this case algorithms based on stabilized Gauss-Netwton methods[5,6,7] can be used to solve iteratively the associated minimization problem.

The choise of a polynomial to model data behaviour, often used in metrological applications, is here considered and a devoted iterative algorithm is presented. It takes into account some special features: both the polynomial function to be minimized and the high precision measures to be processed.

2 The TLS formulation: an iterative algorithm

Let (t_i, y_i), i=1,2,...,m be a set of experimental data x_i and z_i the true, but unknown, correspondig values and the measurement errors $\varepsilon_i = x_i - t_i$, $\eta_i = z_i - y_i$

be uncorrelated random variables with 0-mean and equal variance. Consider as function $\Phi(\bullet,\alpha)$ a polynomial of degree $n < m$

$$\Phi(t,\alpha) = p_n(t) = \sum_{i=0}^{n} \alpha_i x^{n-i} \qquad \alpha \in \mathbf{R}^n.$$

The TLS approach requires to find n+m+1 unknown α and $\mathbf{x}$ such that:

$$\| \Phi(\mathbf{x}, \alpha) - \mathbf{y} \|_2^2 + \lambda \| \mathbf{t} - \mathbf{x} \|_2^2 = \min!, \quad \mathbf{y} \in \mathbf{R}^m,\ \mathbf{t} \in \mathbf{R}^m,\ \mathbf{x} \in \mathbf{R}^m \qquad (1)$$

where the constant λ is given proportional to the ratio of the variances of the errors. The solution cannot be obtained through the robust SVD algorithm, but only solving the so-called *orthogonal distance problem* (1) with iterative algorithms that can be unstable and not efficient for problem of large size.

The proposed algorithm comes from the following considerations. The solution to problem (1), if there exists, must satisfy the nonlinear system of n + m + 1 non linear equations:

$$\frac{1}{2}\frac{\partial F(\mathbf{x},\alpha)}{\partial x_j} = (p_n(x_j) - y_j)\, p_n'(x_j) + \lambda(x_j - t_j) = 0, \qquad j=1, \ldots, m \qquad (2')$$

$$\frac{1}{2}\frac{\partial F(\mathbf{x},\alpha)}{\partial \alpha_k} = \sum_{i=1}^{m} (p_n(x_i) - y_i)\, x_i^{n-k} = 0, \qquad k= 0, \ldots, n \qquad (2'')$$

For any fixed α, each equation in Eq. 2' can be considered a polynomial equation of degree 2n-1 in the unknown x_j, therefore it has at least one real zero. For any fixed $\mathbf{x}$, Eq. 2'' are the usual normal equations arising in the least squares fitting to data (x_i, y_i), i = 1,... , n. Thus finding polynomial roots and solving an LS problem are the body of the iterations:

0. Construct the starting point $(\mathbf{x}^{(0)}, \alpha^{(0)})^T$
1. Repeat until the stop conditions are satisfied:
   1.1 $\forall$ j = 1, ... , m, find all the roots of Eq. 2' for $\alpha = \alpha^{(k-1)}$ and define $x_j^{(k)}$ as the real root closest to t_j;
   1.2 find $\alpha^{(k)}$ by solving the normal Eqs. 2'' for $x_j = x_j^{(k)}$.

As to the starting point $(\mathbf{x}^{(0)}, \alpha^{(0)})^T$, $\mathbf{x}^{(0)} = \mathbf{t}$ is assumed and consequentely the starting vector $\alpha^{(0)}$ is obtained by computing the standard LS solution, since obviously if $\mathbf{x} = \mathbf{t}$ the TLS problem turns to be the classical linear LS problem.

Moreover it is reasonable to required that the expected value for x_j differs from t_j no more than the precision of the measures, due to their high quality. This latter remark is considered firstly to chose $x_j^{(k)}$ in the step 1.1, secondly to stop the procedure if, for some j, $|x_j^{(k)} - t_j|$ becomes greater than a preassigned value depending on the measures precision. This is a stop condition that must be added to the criteria that are commonly used in minimization algorithms:

- maximun number of iterations,
- norm of the difference vector between two successive approximations of $\mathbf{x}$ and α,
- decreasing behaviour of $F(\mathbf{x},\alpha)$.

Finally it has to be underlined that in both steps, problems at most of order 2n-1 are to be solved, where in practical situations n is small (i.e. < 6), while the usual minimization algorithms[5,7] work in a (n+m+1) dimensional space.

3 Application to metrological data

The algorithm has been used to process data obtained from Istituto di Metrologia "G.Colonnetti", CNR, Torino. The data describe the melting plateau of pure substances, where the independent variable is the inverse of the melted fraction F of the substance sample and the dependent variable is the variation of temperature. The extrapolated value to $F = 1$ must be accuratelly computed. In each experiment the number of data is always small because of the high cost of the experiment itself ($m < 10$) and the errors can be modelled as required in the TLS formulation. For physical reasons a linear model would not be sufficient to represent the data; a second degree polynomial ($n = 2$) is generally sufficient as an empirical model.

Several datasets have been analyzed and the minimum problem in Eq.1 has been solved by means of the proposed algorithm and also by means of the NAG routine E04KDF, based on a modified Newton-type method[6].

As general remark, the algorithms find approximated values for $\mathbf{x}$ and α that do not met all the conditions to be a minimum, but a lower point could not be found, in fact the algorithms find the minimum as far as the accuracy of the machine permits. Moreover the CPU times are equal in most cases.

In Table 1 the results obtained using the two algorithms are compared for two different datasets: $m = 9$, $\lambda = 1$, on the left side and $m = 6$, $\lambda = 0.3$, on the right side. In columns 1 and 4 the differences between the solution provided by the presented algorithm and the one by NAG algorithm are reported. The last three rows correspond to the polynomial coefficients, that are $\alpha \cong (.005773,-.81951, 89.79203)^T$ for the first example and $\alpha \cong (.029564,-.34633, .93.59288)^T$ for the second one. Gradient components are also shown. The CPU times are 0.08s and 0.04s for the present algorithm, 0.07s and 0.04s for the NAG routine, respectively in the two examples.

It has to be noted that the main difference can be found in the gradient components, that are the derivatives with respect to the polynomial coefficients, according to the step 1.2.

The LS and TLS solutions are displayed in Figure 1, which refers to the example in the right side of Table 1 ($m = 6$ and $\lambda = 0.3$).

Table 1: Comparison between algorithms (the differences of the polynomial coefficients in bold)

Difference	grad/Alg	grad/NAG	Difference	grad/Alg	grad/NAG
-.154-05	-.22-05	-.10-06	-.390-06	-.16-06	-.44-08
.344-05	.53-05	-.14-07	.856-06	.37-06	.80-09
.226-05	.37-05	-.11-07	.723-06	.32-06	-.21-08
.120-05	.20-05	.14-07	.101-05	.46-06	15-08
-.690-08	-.10-07	.30-08	-.461-06	-.21-06	.11-07
-.109-05	-.18-05	-.21-07	-.273-05	-.14-05	-.51-08
-.153-05	-.26-05	-.10-07			
-.193-05	-.33-05	-.25-07			
-.195-05	-.33-05	.11-07			
-.712-07	-.15-12	-.86-05	**-.330-06**	.15-11	-.63-07
.140-05	-.34-12	-.29-06	**.209-05**	29-12	-.85-08
-.339-05	-.17-12	.76-07	**-.272-05**	-.11-12	27-08

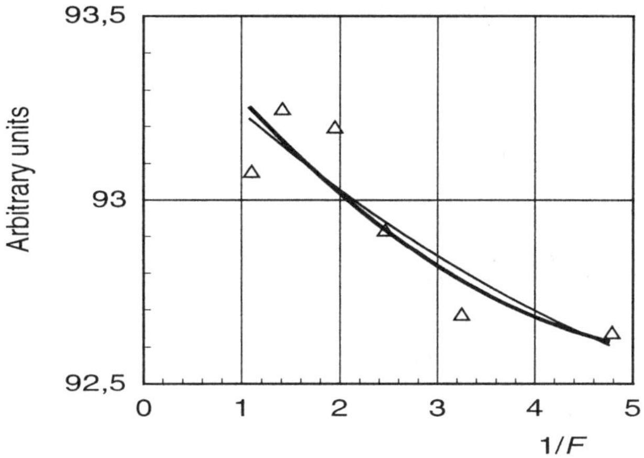

Figure 1: Comparison between TLS (thick line) and LS solutions.
The original data (m = 6) are also shown. F = melted fraction.

References

1. G. Golub, C.F. Van Loan, *SIAM J. Num. Anal.* **17** (1980)
2. G. Golub, in *Advanced Mathematical Tools in Metrology II*, Ciarlini, Cox, Pavese, Richter Eds, World Scientific, 48 (1996)
3. S. Van Huffel and J. Vanderwalle, *SIAM*, Philadelphia (1991)
4. A. Premoli *et al*, in *Advanced Mathematical Tools in Metrology II*, Ciarlini, Cox, Pavese, Richter Eds, World Scientific, 206 (1996)
5. J.J. Moré, in *Numerical Analysis*, G.A.Watson Ed, Lecture Notes in Mathematics **630**, Springer Verlag, Berlin (1977) p105-116.
6. P.E. Gill *et al, Practical optimization.* Academic Press, London (1981)
7. H. Schwetlick and V. Tiller, *Technometrics*, **27**, n.1, 17 (1985)

Advanced Mathematical Tools in Metrology III
Edited by P. Ciarlini, M.G. Cox, F. Pavese & D. Richter

DESIGNING AXIALLY SYMMETRIC ELECTROMECHANICAL SYSTEMS OF SUPERCONDUCTING MAGNETIC LEVITATION IN MATLAB ENVIRONMENT

E.FRANTSUZ, V.KHAVINSON

D.I.Mendeleyev Institute of Metrology (VNIIM)
19 Moskovsky Pr., St. Petersburg 198005, Russia

F.SHIOTA, Y.MIKI, R.MOSSAVATI

National Research Laboratory of Metrology (NRLM)
Tsukuba, Ibaraki 305, Japan

Abstract

A method for forward estimating the shape of an electromechanical system of superconducting magnetic levitation in Matlab environment is suggested.

1 Introduction

We use superconducting magnetic levitation for equating electrical and mechanical energies while a mass in the Meissner state is moved vertically via a superconducting current-carrying-coil [1]. The thought experiments with an ideal electromechanical system aim to find one term of equation from the results of combined measurements of the others. As the measurements should be performed with the aid of a real system while the levitated body rests at one of several positions over the range of its vertical displacement, the shape and design of this system has a strong influence on the attainable uncertainty. Therefore it is important to have a tool for forward estimating the shape of the electromechanical system under consideration. The work suggests the program facilities for a designer to evaluate the following characteristics for any system of given shape:
- the vector components of the equilibrium trajectory, that is, the values of ma gnetic flux, coil current, and vertical position of the levitated body;
- the magnetic induction map over the system cross-section;
- the terms describing oscillations of the levitated body for both the constant- current and constant-magnetic-flux modes, namely, the oscillation frequencies and frequency corrections due to system non-linearity.

2 Features of the Method

In the case of axially symmetric electromechanical systems, it is reasonable to develop the approach of Todd I. Smith [2] for calculating magnetic fields due to a coaxial aggregate of a solenoid and a superconducting cylinder. The idea consists of approximating the levitated body, which is an axially symmetric superconducting shell with continuous current density distribution along a generatrix, by a number of perfectly conducting rings with uniform current density over each ring. In calculations, this allows the replacement of integrals by finite sums and the boundary condition that the trapped magnetic flux must be constant through the superconducting shell leads to a linear system of equations. The solution for the system yields the ring current values and they in turn allow the determination of the field at any point provided that the point does not lie within a few ring spacings from the shell.

The known expressions for the vector potential of the single circular current loop [3] and for the mutual inductance between two axial rings [4] are used as the basis for calculations. The distance between adjacent rings is chosen to be a compromise between the resulting errors and the required time for calculation. Irregular division with the smallest spacings at the edges of the superconducting shell where the current density changes rapidly, proved to be optimum.

To find the levitation force, one needs to integrate the magnetic pressure over the superconducting shell surface area. The pressure is governed by the magnetic ind uction near each surface element i.e. in the region where uncertainties of this method are too large. To overcome the problem, we assume that the ring current values found represent the actual linear screening current density at each point, when divided by the length of the division interval including the point. Hence the current density at any point of the superconducting shell generatrix is straightforward to find by interpolation. To calculate the true value of magnetic induction, the initial current rings near the point under consideration are replaced by the req uired number of rings, the current in each ring being determined by the local current density.

3 Sequence of calculations

In cylindrical coordinates with the axis of the levitating system vertical.

1. Determine the break sequence of the superconducting shell by the horizontal plane set of equivalent ring locations.
2. Determine the matrix of mutual inductances for a set of equivalent rings.

3. Find the portions of the total vector potential due to the levitating coil as the superconducting shell is levitated at the vertical position z.
4. Find the screening current density distribution along the shell generatrix as a function of z.
5. Find the magnetic induction distribution near the shell surface and levitation force as a function of z .
6. From the above function calculate the terms describing oscillations of the levitated body [5].
Effective use of the program tools that MATLAB environment makes available appears to be quite adequate to solve the problem and the above calculations do not require a lot of computing power.

Acknowledgments

R. Mossavati acknowledges the support of the Science and Technology Agency of Japan through a STA Fellowship.

References

[1] D.B.Sullivan and N.V.Frederick, *IEEE Trans. Magn.*, **MAG-13**, 396 (1977).

[2] T.I.Smith, *J.Appl.Phys.*, **44**, 852 (1973).

[3] J.D.Jackson *Classical Electrodynamics*, (Wiley, New York, 1962).

[4] P.L.Kalantarov and L.A.Tseitlin in *Calculation of Inductance*,(3rd ed., Leningrad, 1986), (in Russian).

[5] E.T.Frantsuz *et al*, *Metrologia*, **33**, To be published (1996).

Advanced Mathematical Tools in Metrology III
Edited by P. Ciarlini, M.G. Cox, F. Pavese & D. Richter

DATA FLOW EVALUATION IN METROLOGY

N. GREIF, H. SCHREPF

Physikalisch-Technische Bundesanstalt,
Abbestr. 2-12, 10587 Berlin, Germany

Abstract

The project shortly introduced below dealt with data flow analysis for software-controlled measuring instruments subject to legal control. To support data flow evaluation a software tool was developed.

1 The Problem

Software-controlled measuring instruments are being increasingly used in legal metrology. Often a software program is the key element of a measuring device. As a part of the device, the software has to be evaluated in the framework of a type approval procedure, too. The use of software in legal metrology requires compliance with additional quality criteria. Such additional software requirements are for example safe data handling, systematic software tests, program maintainability, and the protection of the software against manipulation and fraudulent use. The last item has forced the use of sealing means. The software has to be protected by mechanical (lead seal) or electronic sealing (checksums). Since the provider or the user of the device is usually allowed to customize the software, the protection can cover only parts of the program. Imagine for instance a butcher's weighing machine, showing not only the weight of a sausage, but the price per pound and the total price. The butcher has to change the internal list of prices from time to time. So the software cannot be sealed as a whole. On the software-oriented level this means that a clear separation of the software into parts subject to legal control and parts not subject to legal control has to be achieved.

Each software program can be considered as a network of data and their interrelations. Each value computed by the software depends on other values (variable values, constants, return values, ...). Each statement of the code establishes a number of edges between a new (variable) value and those data, which are involved in building it. Some of these computations result in an output to the user. The proper device output (the weight for a weighing machine, the voltage for a voltmeter, etc.) is subject to legal control, and so are all data which contribute to its computation. Thus, for the type approval procedure the software is divided into that part, which contributes to a value being subject to legal control and those parts, which do not contribute. This separation can only be done successfully, if the complete data flow of the software is known.

2 The Solution Developed

Today, finding out the data flow is a time-consuming task carried out manually. No commercial software tool is available to analyze the data flow in the required manner. To improve the situation, a tool for extracting, analyzing, and visualizing the data flow of metrological software was developed. The application of the tool provides a data flow diagram including all interconnections of symbol values (variable values, constants, return values, ...). The way of constructing the value of certain variables and the influences of other symbols in this process are perceptible. For more details about data flow diagrams see [1].

Preparing and performing the data flow analysis the following steps are necessary:

1. Parse the software under test and collect all basic information. Basic information is for example variables and constants used in the program, number and names of all routines, location of read and write accesses for all variables.

2. Combine basic information to collections. Collections needed for analysis are for example the call hierarchy, or the list of all formal and actual parameter pairs.

3. Do the analysis: Start at the output value which is of special interest, find all symbols which are involved in making this value and repeat analysis for them recursively.

4. Prepare the graphical output. List all routines and global symbols which should be subject to legal control.

The first step is a problem common to all kinds of software analysis. To realize this step a commercial tool is used. Since the tool runs on VMS machines, we selected the DEC Source Code Analyzer. The other steps were newly implemented. The tool has a Motif interface and runs in interactive or in batch mode. The tool architecture enables analyses for several languages, but up to now the development was made only for programs written in C.

3 Current Restrictions

The steps mentioned above are only a rough outline. There are several major problems connected with data flow analyses:

1. Symbol transformations: Each time the normal flow of program execution enters a new routine or leaves a routine, local symbols have to be replaced. Formal parameters are replaced by actual parameters and vice versa, local symbols given back are replaced by the routine name carrying a certain value.

2. Statement sequence: If the value of a certain variable x is of interest, we have to identify that statement, which is responsible for this value. We are looking for the last assignment to x which is not overwritten by another. To find it, we have to evaluate the 'order' of statements. The order of execution does not always correspond to lexical order in source code. Although they differ in certain cases the tool in its current version considers lexical order to find the last relevant assignment to a variable. This is one of the current restrictions of the tool.

3. Symbol references: To change a value in memory a programmer usually uses the name of the memory cell (a variable or constant name). In language C a programmer has several ways to avoid using the variable name. He may install new names for the same memory object or even use its address only. The tool finds most of the aliases of a variable, but it cannot find all. Restrictions in this field are for instance fourfold redirections or addresses (pointers) evaluated by arithmetic operations.

To make it possible to recognize variables which influence the data flow conditionally, the tool must be supplemented by a control flow analysis component.

Reference:

[1] N. Greif, H. Schrepf: Data Flow Analysis in Legal Metrology, in: M. Ross, C.A. Brebbia, G. Staples, J. Stapelton (eds.), Measuring and Maintaining Quality, Proc. 3rd Int. Conf. on Software Quality Management, Seville 1995, vol. 2, 265-276, Computational Mechanics Publications, 1995.

Advanced Mathematical Tools in Metrology III
Edited by P. Ciarlini, M.G. Cox, F. Pavese & D. Richter

A GENERALIZED DATA MODEL FOR INTEGRATING CLINICAL DATA AND BIOSIGNAL RECORDS OF PATIENTS

H.GROSS, V.HARTMANN
Physikalish-Technische Bundesanstalt, Abbestrasse 2-12, D-10587 Berlin, Germany

1 Introduction

The progresses in biomedical sensoring, especially the trend towards device miniaturisation and distributed instrumentation, results in a continuously increasing amount of measured data. Often multichannel sensor- and data-acquisition systems are used to realise a measuring with high sampling frequencies over different time periods and under different environmental conditions. Long data or signal records appear describing the time course of special parameters or signals. Especially in bioelectrical and biomagnetical research, this is a current challenge.

The measured data should be accessible for different users, i.e. researchers of the same or of other institutions or employees of industrial companies in order to allow further processing and evaluations.

The completeness of documentation implies that apart from the measured data records a variety of additional information has to be stored. One group of these information describes the measuring devices, such as the technical description of the sensor system with the connected datastorage unit (A/D-converter and computer system). A further group of information characterises the measured object itself, i.e. all further information describing the current state of the patient (clinical data).

The *significance* of each measured data record or data, the *completeness* of the data and their *accessibility* imply the need for advanced data management tools, that is modern database management systems (DBMS).

2 Data models and basic methods

Today relational databases have a dominant position over their predecessors based on hierarchical or network data models. The relational data model provides several important advantages:

- *Data independence*, that is the data representation is independent of the interface to the application. The physical representation can be modified without affecting application programs.
- *Declarative manipulation*, i.e. a standardised structured query language (SQL) is available for creating and manipulating databases. It allows the DBMS to choose from alternative mechanisms and to obtain the desired results from the physical realisation of the databases.

• *No redundancy*, if the process of normalisation is applied to the relational model. In practice, this means a structuring of the relations without repetitions in their non-key attributes.

Furthermore, the major thrust in database technology is concentrated on enhancing the functionality and the performance of relational systems. The realisation of distributed relational databases on different nodes in a computer network is an example for more functionality and the possibility to optimise the usage of available resources. A second example is the capability to handle the BLOB (binary large objects) data type within relations. BLOBs can be the digitised data records of any technical instrumentation such as an electrocardiographic recording system.

The basic method for the application of a DBMS is the creation of the *conceptual model* for the data that should be handled. That means the appropriate entity-relationship model in the case of a relational system has to be developed. Once this important step is done, the conceptual data model has to be translated to the *physical model* of the target DBMS (e.g. Oracle, Rdb, Sybase, SQLServer). During this transformation the normalisation of the relations is carried out, i.e. additional tables can be created. For both steps modern database systems offer special tools with integrated graphical editors. Some CASE-tools allow the design of the conceptual model independently of the target DBMS. For the physical model they support the translation to almost all relational database systems on the market.

3 Functional requirements

If we focus to the issue of a database for ECG and MCG records and all further relevant patient data, we have to integrate biosignal records and standardised clinical data. The records are measured with a special instrumentation which guarantees undisturbed signals of highest quality. These records originate from clinical patients, who suffer from cardiac diseases. Their diagnosis are confirmed by standardised clinical investigations and the expertise of experienced cardiologists.

The functional requirements have to account for the demands of the bioelectrical and -magnetical research laboratories from the daily work and for the demands resulting from the intensive co-operation with clinical departments where the patients are treated. To be effective, the database design addresses the following:

• The complete process of data processing from acquisition to evaluation or interpretation is supported, i.e. the current processing state of each biosignal record and his history is documented.

• The clinical terms is coded uniquely and uniformly for the reliable retrieval.

• The design has to be extensible so that new types of data (resulting from new algorithms or new clinical investigations) can be added without necessitating changes to the definitions of existing database tables.

• A flexible design has to allow the represention of data which consist of heterogeneous data types along with complex data such as dynamic parameter vectors or large blocks of binary data.

4 Results

These requirements are satisfied by the design decisions for the data model:

- clinical findings and MCG or ECG-recordings with their protocol values are documented in the table **Events** which have as columns the datetime and a pointer to the table **Methods**, where the acquisition and evaluation methods are defined;
- each of the entries in table **Methods** is defined by a vector from the table **Method_Parameters** whose parameter identifiers reference to the table **Parameters**;
- the results of a event (i.e. the application of a method), are saved in the datatype specific tables **P_Numbers**, **P_BLOBS**, **P_Strings**, **P_Booleans** and **P_Datetimes** which are special coupling tables between the tables **Events** and **Parameters**.

Figure 1 presents the main part of the normalised table schema resulting from the entity relationship diagram of the data model, if the database server is Oracle.

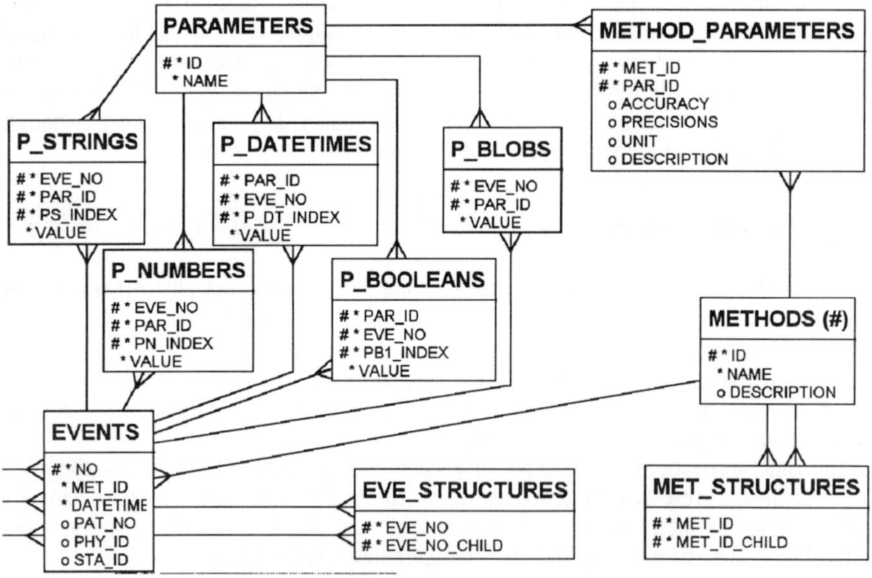

Figure 1: Part of the table data schema of the data model

Advanced Mathematical Tools in Metrology III
Edited by P. Ciarlini, M.G. Cox, F. Pavese & D. Richter

ASSESSMENT OF THREE-DIMENSIONAL STRUCTURES IN CLINICAL DENTISTRY

V. JOVANOVSKI, L. ZOU, W. M. TAY, E. LYNCH
St Bartholomew's and the Royal London School of Medicine and Dentistry, London, UK

M. G. COX
National Physical Laboratory, Teddington, UK.

1 Introduction

Many procedures in dentistry involve assessment of the morphology of teeth and other oral structures. Since much of this assessment is subjective, it is of benefit to researchers and clinicians if they can be provided with objective quantitative information in a form which permits accurate measurements and comparisons of morphological features. This can be brought about by digitising tooth surfaces as sets of (x, y, z) co-ordinates and constructing a mathematical model with the aid of appropriate supporting software.

2 Materials and Methods

In our current studies, tooth surfaces are replicated *in vivo* using a standard polyvinylsiloxane impression material. The replicas are digitised by a Ferranti Merlin co-ordinate measuring machine (CMM) fitted with an infra-red laser probe (Renishaw OP2, $\lambda = 830$ nm). A typical data set consists of approximately 40 000 points sampled over a 12 mm x 12 mm region with an accuracy of 5 μm (RMS). The acquisition time is 20 minutes per replica, with a batch mode permitting unattended processing of up to 36 replicas.

In order to compute length, angle, area and volume or to visualise the surface, we need the ability to determine the z co-ordinate and the surface normals for any values of x and y. This requires that the data set be represented in the form $z = f(x, y)$. It was concluded from previous experience that the most suitable representation would be obtained using bivariate polynomial splines.

A certain degree of complication is introduced by the fact that the (x, y) co-ordinates of the digitised points do not form a rectangular grid. Ideally, the optical probe moves across the surface of the replica in parallel lines but the nominally constant y co-ordinate cannot be perfectly controlled, and the x-spacing of points along each line is also variable. The set of points $\mathbf{p}_i$ representing a particular surface is first converted to one which lies on a rectangular x-y grid. Although several methods exist for the interpolation of general scattered-data sets, in this case use can be made of the fact that for each fixed index i the points $\mathbf{p}_{ij}$ lie on a nearly-planar spatial curve. For each i a spline curve interpolant passing through the points $\mathbf{p}_{ij}$ is constructed and evaluated on a set of uniformly spaced values of x, producing a set of points $\mathbf{q}_{ij}$ whose x co-ordinate is constant for each index j. Then, for each j a spline curve interpolant passing through the points $\mathbf{q}_{ij}$ is constructed and evaluated on a set of uniformly spaced values of y, producing a set of points $\mathbf{r}_{ij}$ whose z co-ordinates approximate the height of the tooth surface at the vertices of a rectangular grid defined by the uniformly spaced x and y values. Finally, the points $\mathbf{r}_{ij}$ are interpolated with a bicubic spline surface.

This representation allows efficient implementation of measurement-related functions such as grey-scale shaded display from any chosen viewpoint, generating a profile by intersecting the surface with a plane of arbitrary orientation and measurement of lengths and angles on such profiles.

Comparison of two sequential measurements of the same replica can reveal any changes over time, such as those due to decay, plaque formation and restorative work. For this, the two data sets need to be in the same frame of reference in the sense that corresponding points (those which represent the same point on the surface) should have the same co-ordinates. However, co-ordinate data from any two replicas will not generally lie in the same reference frame. The major contributing factors are variations in orientations of the replicating assembly and of the replicas when placed on the measuring platform of the CMM. Attempting to reduce these variations by physically constraining the positioning and orientation is impractical as it would still permit errors which are orders of magnitude greater that the CMM measurement accuracy. The reorientation can instead be applied to the co-ordinate data sets by transformations which are the equivalents of physical translations and rotations by determining the parameters of an optimal rigid 3-D transformation $\mathbf{x} \mapsto \mathbf{R}\mathbf{x} + \mathbf{v}$ where $\mathbf{R}$ is a proper orthogonal matrix and $\mathbf{v}$ a translation vector. Theoretically, the optimum transformation can be found by identifying at least three pairs of corresponding points. Such 'landmark identification' methods are in widespread use, but they are inappropriate if the surface is smooth and relatively featureless, as tooth surfaces often are. Artificial landmarks attached to the tooth are ruled out by the requirement that they would have to remain in place for long periods causing possible discomfort to the patient. Instead, use is made of the facts that (a) some regions of the tooth are likely to remain unchanged and (b) the surface is represented as $z = f(x, y)$ with the partial derivatives of f available.

Software for superposition was written which displays a pair of surfaces and allows the operator to mark the unchanged regions interactively. From these regions a set of representative points on the first surface is generated automatically. These points are fitted onto the second surface by a least squares procedure which determines the parameters of the optimum 3-D transformation.

In order to allow researchers to perform their own analyses the capabilities described above have been incorporated into a user-friendly software package which runs on PC-compatible computers. Additionally, it allows simultaneous measurements of a sequence of superposed surfaces to determine changes in profile and volume. Changes can be visualised by the production of colour-coded difference maps. The results can be exported in standard spreadsheet, word-processing and graphics formats.

3 Results and Conclusion

The work has demonstrated that the assessment of three-dimensional structures in clinical dentistry can benefit from the application of sound mathematical techniques to measured data representing the surfaces of tooth replicas. The overall accuracy of the results is limited by each stage of processing including tooth replication, digitising, interpolation and superposition. To determine this accuracy, a single tooth was replicated five times. The replicas were digitised and superposed into a common reference frame. A set of 1000 points was defined in the x-y plane and the variations in the z co-ordinates were evaluated at each of these points. The RMS variation was found to be 10 μm.

Pilot studies have been carried out involving measurement of plaque thickness, tooth wear and the evaluation of margins of crown preparations, each involving around 80 patients. These initial investigations have shown the system to be reliable and sufficiently accurate for current requirements.

Although the current system is straightforward to use and permits reasonable numbers of replicas to be processed, there will be further benefits to researchers and for educational purposes by upgrading the system hardware to give even greater measurement accuracy and to process greater numbers of replicas. The mathematical techniques developed and the supporting software are such that they will be immediately applicable to such improved systems. In particular, it is expected that the increased measurement accuracy will be carried through the interpolation and superposition stages to yield comparably improved results.

Advanced Mathematical Tools in Metrology III
Edited by P. Ciarlini, M.G. Cox, F. Pavese & D. Richter

MAXIMUM ENTROPY & BAYESIAN APPROACHES TO PARAMETER ESTIMATION IN MASS METROLOGY

THOMAS S. LEAHY
School of Physical Sciences, Dublin City University, Dublin 9, Ireland
Email: tsl@physics.dcu.ie

EXTENDED ABSTRACT

Data analysis models should be both mathematically sound and physically relevant. Consistency and homogeneity should also be characteristics of the analysis. The treatment of measurement uncertainties is critical to the accuracy of the method developed. The logic of Classical Probability theory provides the most appropriate approach to dealing with data of a non-deterministic nature as it is a method of reasoning in the absence of certainty[1,2]. A Bayesian view of probability is adopted, based on desiderata of consistency—namely that the same conclusion should be reached, irrespective of the intermediate routes of evaluation, that all known and available information must be incorporated into the analysis and that equivalent states of knowledge must be described by equivalent mathematical statements[1,3]. From this foundation a complete probability analysis can be constructed.

A thorough practical basis for implementing a measurement analysis strategy has been presented in the ISO Guide[4], now widely accepted. One of its key aspects is a uniform and consistent treatment of all influence quantities. Establishing variance measures for all elements is achieved by assigning Degrees of Belief which reflect the extent of available knowledge about each parameter. This will include all relevant information, but only that information which is available about the parameter[1,2,5].

The Principle of Maximum Entropy (MaxEnt) ensures that Degrees of Belief can be assigned in a consistent manner, free from possible bias. The entropy of a probability distribution gives a measure of the amount of "uncertainty" it represents. Maximising this subject to the constraints of whatever *is* known about the parameter generates the most honest probability assignment since it assumes the least knowledge about the parameter while reproducing any known features. For example, maximising the entropy tells us, as we would expect, that a uniform distribution is the best assignment when nothing is known except the range of values the parameter could adopt. Also when an estimated mean and variance are known, MaxEnt indicates a Gaussian Distribution to be the least subjective.

The analysis of mass calibration data involves the parameterising of an experimental situation and subsequent data reduction of a set of over-determined comparisons to evaluate the mass values of a set of standards[6,8]. A Unified Approach is desired to all aspects of this work, which will ensure optimal estimation and full use of all available information in a non-biased manner[7]. This is done using the principles of consistent reasoning of the ISO Guide, MaxEnt and Bayesian Parameter Estimation.

Statistical Adjustment by the Bayesian Method[6,7,8] ensures the process of consistent reasoning is maintained. It utilises all the available information while not assuming unknown data. It results in parameter estimates of improved accuracy, often providing error-detection and even error-correction, should there be inconsistencies or systematic errors in the data.

References

1. Jaynes, E. T. "*Probability Theory: The Logic of Science*"; Fragmentary Edition, March, 1996, available on the WWW.
2. Garrett, A. J. M. "*Macroirreversibility & microreversibility reconciled: the second law*" pp 139 - 170 in *Maximum Entropy in Action*, B. Buck & V. A. Macaulay, eds. Oxford, 1991.
3. Sivia, D. S. "*Data Analysis: A Bayesian Tutorial*", Oxford, 1996
4 I.S.O. "*Guide to the Expression of Uncertainty in Measurement*", 1993
5 Weise, K. & Wöger, W. "*A Bayesian theory of Measurement Uncertainty*"; Meas. Sci. Tech. **3**, 1992, pp 1 - 11.
6 Leahy, T. S. "*An analysis of Statistical Estimation Methods used in High Accuracy Mass Determination*", pp 259 - 266 in *Advanced Mathematical Tools in Metrology*, P. Ciarlini *et al*, eds. World Scientific, Singapore, 1996.
7 Leahy, T. S. "*An Analysis of Statistical Estimation Methods & Model Parameterisation in Mass Calibration*", Postgraduate Research Thesis, in preparation.
8 Bich, W. "*Bias & Optimal Linear Estimation in Comparison Calibrations*", Metrologia, **29**, 1992, pp 15 - 12.

Advanced Mathematical Tools in Metrology III
Edited by P. Ciarlini, M.G. Cox, F. Pavese & D. Richter

AMPLITUDE AND PHASE DETERMINATION OF SINUSOIDAL VIBRATION IN THE NANOMETER RANGE USING QUADRATURE SIGNALS

A. LINK
Physikalisch-Technische Bundesanstalt
Abbestrasse 2-12 ,10587 Berlin, Germany

For the accurate measurement of the amplitude and phase angle of a sinusoidal vibration, laser interferometry is applied. The paper presents a method for measuring the amplitude and phase angle of a sinusoidal vibration in the nanometer range by using quadrature signals generated by digital signal processing.

1 Introduction

The proposed method makes use of two heterodyne signals which may be taken from a commercial laser vibrometer containing an acousto-optical modulator. The signals chosen are one signal from the interferometer (phase-modulated by the sinusoidal vibration displacement) and the internal oscillator signal. These two signals are synchronously sampled using two-channel A/D converters. The two sampled signals are multiplied and filtered to obtain two quadrature signals (QSs) located in their basic frequency band. The application to these QSs of an arctan function algorithm with phase unwrapping allows the discrete-time phase signal of corresponding displacement values to be generated. The approximation of this phase sequence by a sine function employing least squares estimation (LSE) furnishes the amplitude and the phase of the displacement.

2 Generation of quadrature signals

The quadrature outputs of the homodyne interferometer furnished by two photodetectors provide two output signals with a phase shift of approximately $\pi/2$. Unequal gain in the detector channels and non-zero offset are errors which limit the accuracy of vibration displacement measurements [1]. To avoid an optical generation of QSs, the output signals of a heterodyne interferometer, one signal modulated by the vibration and the oscillator signal are utilized. After A/D conversion, the two digitised signals are used to generate the QSs $u(n) = U\cos\varphi(n)$ and $v(n) = U\sin[\varphi(n) + \rho]$ [2], where n is an integer which denotes the nth sample of the sequencies. The phase signal $\varphi(n)$ contains the modulation component $\varphi_m \cos(\omega_s n + \varphi_s)$ wich reflects the vibration with the known radian frequency ω_s.

ρ is a phase shift error caused by time jitter of the sample generator. It can be assumed that in the current state of digital signal processing, such an error is essentially smaller than a phase shift error occurring when quadrature signals are generated by optical means.

3 Amplitude and phase angle estimation and simulations

Applying to the two QSs known digital signal subroutines which realise an arctan function and phase unwrapping, we obtain within the range of observation $n = 1, 2, \ldots, N$ the phase sequence $\alpha(n) = \arctan[v(n)/u(n)]$. To estimate φ_m and φ_s, the signal model $\alpha(n) = a\cos\omega_s n + b\sin\omega_s n + c$ is used, which is linear with respect to the parameters $a = \varphi_m \cos\varphi_s, b = -\varphi_m \sin\varphi_s$ and c which is a constant. On the basis of this model, the parameters a, b and c are calculated using the LSE and from their values, the phase amplitude and phase angle can be estimated using the relationships $\hat{\varphi}_m = (\hat{a}^2 + \hat{b}^2)^{1/2}$ and $\hat{\varphi}_s = \arctan(-\hat{b}/\hat{a})$.

When examining the performance of amplitude and phase estimation, it is necessary to include in this analysis distortions caused by the generation of heterodyne QSs and depending on a finite resolution of A/D conversion. We also take into account that in measurement practice sinusoidal vibrations are superimposed with different noise signals. The analysis has shown that the mean values of the phase amplitude and phase angle deviation caused by these disturbing sources mainly depends on the quantization accuracy of the quadrature signals. Realising a full scale 8-bit quantization of quadrature signals we can determine displacement amplitudes of 1nm with an uncertainty smaller than 0.5%.

4 Conclusion

The proposed signal processing method is based on quadrature signals generated by digital signal processing preventing errors that typically occur when quadrature signals are generated by optical means. The measured vibration signal can be regenerated with high accuracy by applying to the phase sequence $\alpha(n)$ methods designed for discrete-time signal processing. This accuracy depends mainly on the realized quantization accuracy of A/D conversion.

References

1. P. L. M. Heydemann, *Applied Optics*, vol. 20 no. 19, pp.3382-3384, (1981).
2. A. Link *et al*, *Proc. 2nd Int. Conf. on Vibration Measurements by Laser Tech., Ancona, Italy*, pp. 37-48, (1996).

Advanced Mathematical Tools in Metrology III
Edited by P. Ciarlini, M.G. Cox, F. Pavese & D. Richter

A CLASS OF SYMMETRIC COMPACTLY SUPPORTED WAVELETS AND ASSOCIATED DUAL BASES

M.L. LO CASCIO, F. PITOLLI
Department of Me.Mo.Mat., Università "La Sapienza"
via A. Scarpa 16 , I 00161 ROMA

A class of suitable scaling functions, which generate multiresolution analyses, is introduced. The associated semi-orthogonal wavelets are considered and procedures to construct the dual bases are given.

1 Introduction

Many filtering applications use the wavelet transform based on orthonormal wavelets, as the Daubechies ones, which have the disadvantage to be asymmetric and to have not linear phase (with the exception of the Haar basis which, however, is not continuous). Since in some signal processing applications symmetry is desirable, one has to give up the orthogonality and to consider other families of analyzing functions, such as the semi-orthogonal wave lets or the biorthogonal wavelets. For a extended bibliography see, e.g.[1-3].

In this paper we consider multiresolution analyses generated by a class of scaling functions that allow us to introduce symmetric semi-orthogonal wavelets. In Section 2 these functions are presented and a procedure to construct the corresponding dual functions is outlined. In Section 3 some graphs showing the qualitative behaviour of these functions are displayed.

2 Construction of semi-orthogonal wavelets and dual bases

Consider the compactly supported *mask* $\mathbf{a}^{(h)} = \{a_k^{(h)}\}$, depending on a real parameter $h > 1$

$$a_0^{(h)} = a_3^{(h)} = 1/2^h, a_1^{(h)} = a_2^{(h)} = 1 - a_0^{(h)}, \; a_k^{(h)} = 0 \text{ otherwise.} \tag{1}$$

(Observe that the value $h = 2$ leads to the mask of the B-spline of order 3.)

The choice of the mask in Eq. 1 assures that, for any fixed h, there exists[5] a non negative *scaling function* $\varphi_h(x) \in C^0(R)$, which is the unique solution to the *two-scale equation*

$$\varphi_h(x) = \sum_{k=0}^{3} a_k^{(h)} \varphi_h(2x - k). \tag{2}$$

All functions φ_h have support [0,3], are a partition of unity and are symmetric with respect to $x = 3/2$ [5]. Moreover, they have linear phase [1], that is $\hat{\varphi}_h(\omega) = |\hat{\varphi}_h(\omega)|e^{-3i\omega/2}$, where $\hat{\varphi}_h(\omega) = \int_R \varphi_h(x)e^{-i\omega x}dx$ is the Fourier transform of φ_h.

The scaling functions φ_h have been yet considered in a kind of cardinal interpolation problem [2]. Here, the functions φ_h are used to construct semi-orthogonal wavelets, also called pre-wavelets, and related dual bases, taking into account that, it has been proved [3] in a more general case that, for any fixed h, the functions φ_h generate a multiresolution analysis. In particular we consider the minimally supported pre-wavelet, whose support is $[-2, 3]$, which is unique [1] and can be expressed by [4]

$$\psi_h(x) = \sum_{j=-2}^{2} (-1)^j \mu_{j-1}\varphi_h(2x - j), \quad \mu_j = \int_R \varphi_h(2x + j)\varphi_h(x)dx. \quad (3)$$

All functions ψ_h are antisymmetric with respect to $x = 1$, have linear phase [1] and satisfy the semi-orthogonality condition

$$< \psi_h(2^r x - j), \psi_h(2^s x - k) >= 0, \text{ if } r \neq s. \quad (4)$$

In the wave let decomposition one needs also the dual bases.
Given a function θ, its unique dual function $\tilde{\theta}$ is defined by

$$\tilde{\theta} \in span\{\theta(x - k)\}, < \theta(2^r x - j), \tilde{\theta}(2^s x - k) >= \delta_{rs}\delta_{jk}.$$

The dual functions $\tilde{\varphi}_h$ and $\tilde{\psi}_h$ cannot be compactly supported, nonetheless they have exponential decay; moreover, they have linear phase [1].

To construct $\tilde{\varphi}_h$ and $\tilde{\psi}_h$, let us introduce the generalized Euler-Frobenius polynomial associated to a function θ:

$$E_\theta(z) = \sum_{k\in Z} \gamma_k z^k, \quad \gamma_k = \int_R \theta(x)\theta(x + k)dx. \quad (5)$$

It has been shown [1] that the following relations hold:

$$\hat{\tilde{\varphi}}_h(\omega) = \frac{\hat{\varphi}_h(\omega)}{E_{\varphi_h}(z^2)}, \quad \hat{\tilde{\psi}}_h(\omega) = \frac{\hat{\psi}_h(\omega)}{E_{\psi_h}(z^2)}, \quad z = e^{-i\omega/2}. \quad (6)$$

The right hand sides in Eq. 6 can be easily evaluated because, due to the finite support of φ_h and ψ_h, E_{φ_h} and E_{ψ_h} reduce to finite sums and the Fourier transform $\hat{\varphi}_h$ is an integral on a finite interval. The dual functions $\tilde{\varphi}_h$ and

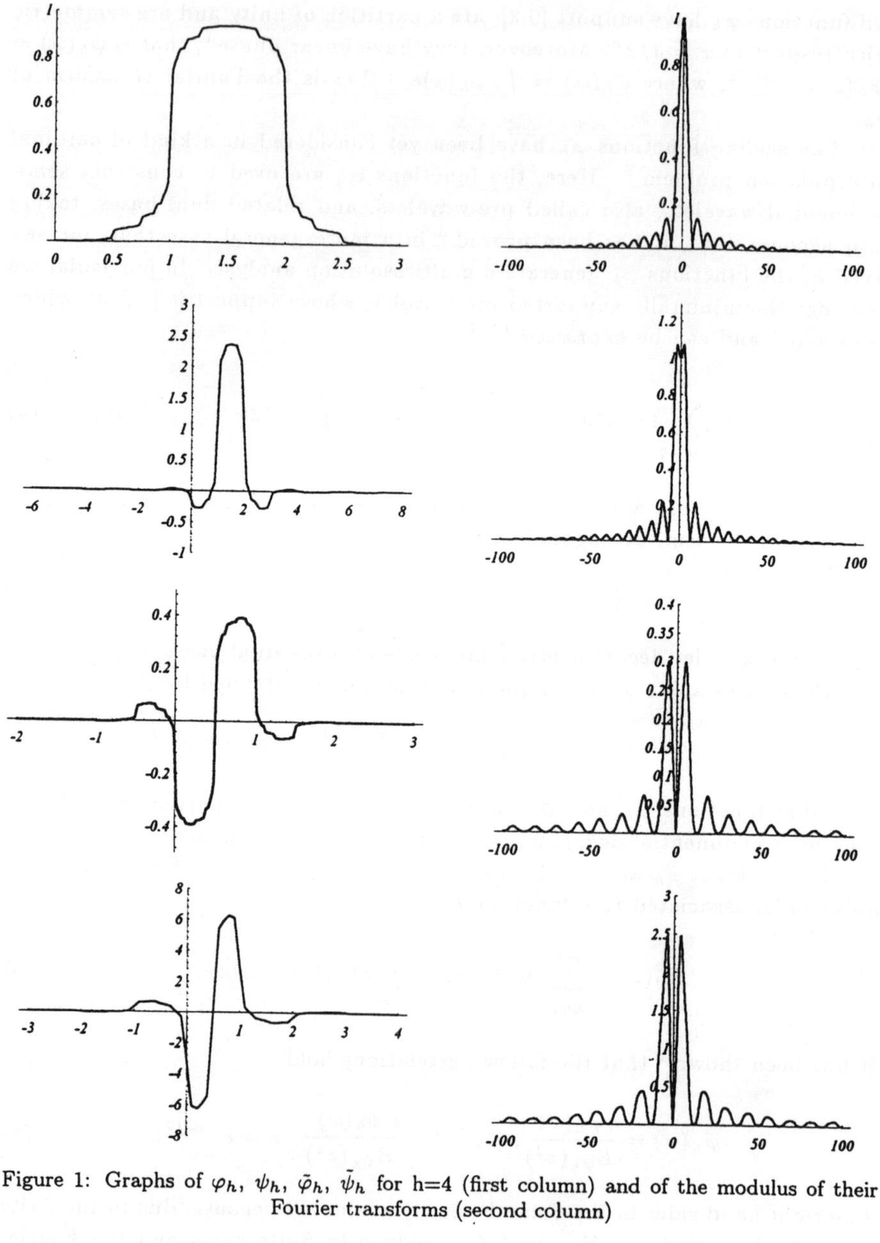

Figure 1: Graphs of φ_h, ψ_h, $\tilde{\varphi}_h$, $\tilde{\psi}_h$ for h=4 (first column) and of the modulus of their Fourier transforms (second column)

$\tilde{\psi}_h$ can be evaluated by the inverse Fourier transform taking also into account that, using Eq. 3, $\hat{\psi}_h$ can be expressed in terms of $\tilde{\varphi}_h(\omega/2)$.

If the wavelet decomposition is required, another procedure has to be followed. Denote by $G_{\varphi_h}(z) = z^4 E_{\varphi_h}(z^2)$ and $G_{\psi_h}(z) = z^8 E_{\psi_h}(z^2)$, that are polynomials of degree 8 and 16 respectively. The application of the inverse Fourier transform in Eq. 6 leads to

$$\tilde{\varphi}_h(x) = \sum_{j\in Z} c_k \varphi_h(x-k), \quad \tilde{\psi}_h(x) = \sum_{j\in Z} d_k \psi_h(x-k) \tag{7}$$

where c_k and d_k are the coefficients of the Laurent series of $1/G_{\varphi_h}$ and $1/G_{\psi_h}$ respectively. The values c_k are given by

$$c_k = -\frac{sign(k)+1}{2}\sum_{|z_j|>1}\frac{A_j}{z_j^{k+1}} - \frac{sign(k)-1}{2}\sum_{|z_j|<1}\frac{B_j}{z_j^{k+1}}$$

where z_j are the zeros of G_{φ_h} and A_j and B_j are the coefficients of the decomposition of the rational function $1/G_{\varphi_h}$. An analogous formula can be derived for the d_k. The coefficients c_k and d_k have exponential decay, thus in practical computation the series in Eq. 7 can be considered as finite sums.

3 Examples

The numerical experiments show that for $h < 3$ the shape of φ_h is similar to the one of the B-spline of order 3. For $h \to \infty$, the shape of φ_h is similar to the one of the Haar function, as expected since $\mathbf{a}^{(\infty)} = (0, 1, 1, 0)$ is just the mask corresponding to it. In Fig. 1 graphs of φ_h, ψ_h, $\tilde{\varphi}_h$ and $\tilde{\psi}_h$ are displayed in the case $h = 4$ together with the corresponding Fourier transforms which give an idea of the analyzing window in the frequency space.

All the integrals involved in the construction of the pre-wavelets and the dual bases, have been carried out with a trapezoidal rule. More efficients algorithms are at present under investigation.

References

1. C.K. Chui *Introduction to wavelets* (Academic Press, New York 1992)
2. M.L. Lo Cascio, F. Pitolli to appear in *Note di Matematica*
3. L. Gori, F. Pitolli *submitted*
4. C.A. Micchelli *Numer. Alg.* **1**, 75 (1991)
5. C.A. Micchelli, H. Prautzsch *Num. Anal.* (Griffith-Watson, eds) Longman Sc. & Tech., Harlow, 192 (1988)

Advanced Mathematical Tools in Metrology III
Edited by P. Ciarlini, M.G. Cox, F. Pavese & D. Richter

ANALYSIS OF SURFACE TOPOGRAPHY BY MAXIMUM ENTROPY POWER SPECTRUM ESTIMATION

G. OGANESIAN
Technical University of Armenia, Terian Str. 105, Jerewan, Armenia
K. HASCHE, H. MOECK
Physikalisch-Technische Bundesanstalt, 10587 Berlin, Abbestr. 2-12, Germany

The maximum entropy method is used to estimate of two-dimensional power spectra. The work is based on an algorithm given by Lim and Malik. The estimation serves the analysis of data obtained in surface topography measurements.

1 Introduction

The recent progress achieved in the 3D analysis of surface topography has increased the interest of researchers and industrial users in the new techniques (scanning force microscopy, scanning electron microscopy, interference optical microscopy combined with CCD camera technique, ...). In this context, the sets of 2D measurement data have to be analyzed with respect to differing practical needs [1]. In [2], an algorithm has been given which approximates surface measurement data by a sum of non-orthogonal trigonometric functions and a set of residual random data. It is known that this algorithm fails in certain situations. This fact has been the reason, though not the most important one, why the work reported here was carried out.
It is an attempt to a more precise frequency analysis.

2 The algorithm

The power spectrum estimation (PSE) based on the maximum entropy (ME) method is one of several possible approaches.

It is the ME philosophy to assume that the process under investigation is as random, or statistically structureless, as possible. When this principle is applied to the PSE problem, we have to select as our spectral estimate a function that maximizes the entropy rate and is consistent with the autocorrelation function (ACF) values of the measurement data set.

An iterative algorithm for solving the ME PSE problem has been suggested by Lim and Malik in [3]. As mentioned by the authors, it is not known theoretically under which conditions the algorithm converges. Nevertheless, it has been empirically observed that the algorithm always converges to the ME PSE when a sufficiently great length of the Discrete Fourier Transform (DFT) and a sufficiently small parameter which stops the iterations are chosen.

The algorithm uses direct and inverse Fast Fourier Transform (FFT) in each iteration. The choice of the ACF size, of the FFT size and of some initial values of parameters influences the results obtained.

3 Applications

The algorithm was implemented in a C language program and run in a Sparc Ultra1 workstation. We have applied the program to the analysis of synthetic data and actually measured data.

In Table 1, three functions used for the generation of synthetic data with the white noise component n(i,j) of the power 1 are listed. Table 2 gives the results obtained for the three examples in dependence on dimensions of ACF and FFT.

Table 1: Synthetic data used on computational experiments

Example 1:	$z(i,j) = \cos(0.6i + 0.44j) + 3\cos(0.9i + 0.66j) + n(i,j)$
Example 2:	$z(i,j) = \cos(0.6i + 0.44j) + 3\cos(1.1i + 1.2j) + n(i,j)$
Example 3:	$z(i,j) = 2.3\cos(0.6i + 0.44j) + 2.1\cos(0.9i + 0.66j) + n(i,j)$
with n(i,j) - white noise of the power 1; data sets with dimensions of 128 x 128	

Table 2: Results of experiments

Example	Dimensions of ACF	Dimensions of DFT	True frequencies	Estimated frequencies
1	25 x 25	256 x 256	(0.66; 0.9)	(0.6627; 0.9081)
1	15 x 15	256 x 256	(0.66; 0.9)	(0.6627; 0.8836)
1	25 x 25	128 x 4096	(0.66; 0.9)	(0.6872; 0.9019)
1	25 x 25	512 x 512	(0.66; 0.9)	(0.6627; 0.8958)
2	15 x 15	4096 x 256	(1.2; 1.1)	(1.1965; 1.0308)
2	15 x 15	256 x 256	(1.2; 1.1)	(1.2026; 1.1045)
2	15 x 15	512 x 512	(1.2; 1.1)	(1.2026; 1.1045)
3	25 x 25	512 x 512	(0.44; 0.6) (0.66; 0.9)	(0.4295; 0.6013) (0.6872; 0.9081)
3	25 x 25	1024 x 1024	(0.44; 0.6) (0.66; 0.9)	(0.4234; 0.5952) (0.6819; 0.9081)

When the two columns "true frequencies" and "estimated frequencies" are compared with the corresponding calculated value, it can be seen that agreement is in general good.

Table 3 refers to the practical example of the calibration of the lateral axes of a scanning force microscope. The measurement data are those of a sinusoidal grating of a well-known periodicity. The ACF was calculated from the corresponding measurement data set "Kalgit". The fundamental frequency which is correlated with the period of the grating was calculated by the TOPO algorithm mentioned above [2] and by ME PSE. It can be seen that the values of the first frequency component agrees satisfactorily and that agreement of the second component is very good.

Table 3: Example for solving the calibration problem mentioned

Dimensions of ACF	Dimensions of DFT	TOPO [2] frequencies	Estimated frequencies
35 x 35	512 x 2048	(0.0204; 0.5903)	(0.0245; 0.5921)

4 Conclusions

Lim's algorithm for the ME PSE was applied. Tests were made with synthetic and real data. On the basis of the results obtained it can be summarized that the application of ME PSE to the calibration of scanning probe microscopes and related problems is conceivable. The efficiency of the algorithm for the analysis of 3D surface topography must be investigated by further evaluation.

Acknowledgements

The authors wish to express their gratitude to Dipl.-Ing. Kerstin Thiele, to Andre Dias (Brazil) and Declan Quinn (Ireland), IAESTE-trainees, and, last but not least, to Houssik Oganesian who studies in Berlin, for their contributions to develop and test the computer program.

References

1. P.M. Leonardo, H. Trumpold, L. De Chiffre, *Annals of the CIRP* (1996).
2. T. Ahbe, K. Hasche, K. Thiele, Proceedings of the XIII IMEKO World Congress, Torino, Vol. 3, 1635 (1994).
3. J.S. Lim, N.A. Malik, IEEE Transactions on Acoustics, Speech, and Signal Processing, Vol. ASSP-29, 401 (1981).

Advanced Mathematical Tools in Metrology III
Edited by P. Ciarlini, M.G. Cox, F. Pavese & D. Richter

INFLUENCE OF DIFFERENT KINDS OF ERRORS ON IMAGING RESULTS IN OPTICAL TOMOGRAPHY[a]

M. ORLT, M. WALZEL, R. MODEL
Physikalisch-Technische Bundesanstalt,
Department of Medical Physics and Information Technology,
Abbestraße 2–12, D-10587 Berlin, Germany

Abstract

An iterative algorithm based on the diffusion model and on a weigted least-squares fit is applied to the inverse problem of optical tomography. Measurement simulations are used to investigate the influence of the FEM error and of data noise on the parameter identification. The results are used for controlling a reasonable relation between the different errors.

1 Reconstruction algorithm

The aim of optical tomography is the imaging of the inner structure (defined by the optical properties absorption μ_a and reduced scattering μ_s') of tissue objects from measurements of reflected and transmitted near-infrared laser light with applications in medical diagnostics.

The inverse imaging algorithm is based on a fit of the simulated data $\boldsymbol{J}^{sim}(\mu_a, \mu_s')$ to the (given) measured time-resolved transmittance $\boldsymbol{J}^{mes}$. The vector $\boldsymbol{J}^{sim}(\mu_a, \mu_s')$ contains the simulated output photon flux, provided $\mu_a(\boldsymbol{x})$ and $\mu_s'(\boldsymbol{x})$ are the actual distributions of the optical parameters. $\boldsymbol{J}^{mes}$ is the vector of the corresponding measurement data for a fixed arrangement of several photon sources, detectors, and times. The reconstruction algorithm consists in the choice of an initial guess and an iterative correction of the optical parameters μ_a and μ_s' until the corresponding simulated vector is well fitted to the measurement vector [3,4]. When a weighted l_2-norm is used, the resulting problem becomes a least-squares-type optimization problem,

$$\|\boldsymbol{J}^{sim}(\mu_a, \mu_s') - \boldsymbol{J}^{mes}\|_{\boldsymbol{w}}^2 = \min! \qquad (1)$$

The weights compensate differences in the orders of magnitude of time-of-flight curves depending on the distance between sources and detectors. The diffusion theory describes the light propagation in scatter-dominated media [2]. In the time-dependent case, it leads to a parabolic differential equation for the photon density with boundary condition of the third kind and a initial condition. The Finite Element Method (FEM) is used to solve the forward problem.

[a]This work was supported by the German Federal Ministry BMBF, grant 13N6307.

In practice, the measured and the simulated data are perturbed by measurement or numerical errors, respectively. Other simulation errors stem from the idealized mathematical model for the light propagation, the photon source and the detector. As the problem is ill-conditioned, these errors have a strong influence on the reconstruction results.

2 Influence of FEM discretization

Here, the influence of the FEM-error resulting from the spatial and time discretization of the object is demonstrated. The imaging problem was restricted to the reconstruction of the absorption coefficient for the following tests. A rectangle 4cm × 4cm in size served as a test object Ω. The absorption $\mu_a(\boldsymbol{x})$ was approximated by piecewise constant functions on a separate rectangular grid with a pixel size of 2.67mm × 2.67mm.

The FEM solution converges to the exact solution, if the discretization parameters $(\Delta x)_{max}$ and $(\Delta t)_{max}$, the maximum node distance and the maximum time step length, tend to zero. A simulation $\boldsymbol{J}^{fine}(\mu_a^{orig})$ for very small $(\Delta x)_{max}$ and $(\Delta t)_{max}$ and a given parameter distribution μ_a^{orig} was used as reference data $\boldsymbol{J}^{mes}$. Using different coarser FEM discretizations for the simulations within the reconstruction algorithm the influence of discretization error on the imaging results is clearly visible (see Figure 1).

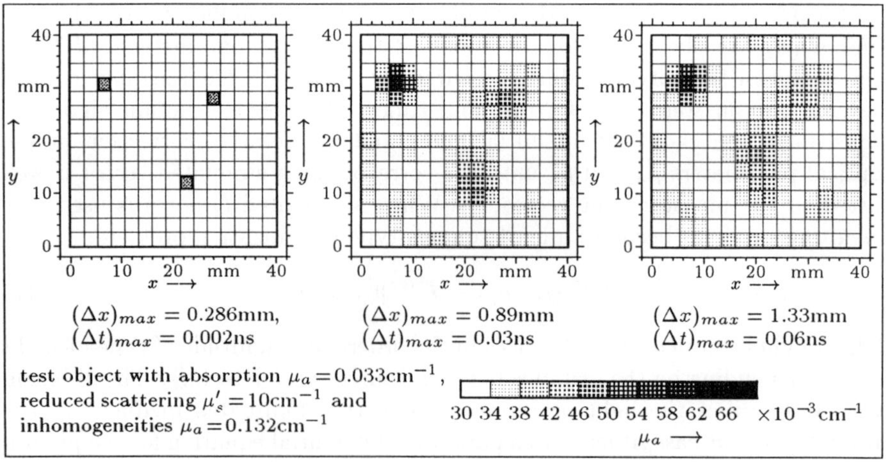

Figure 1: Test object and reconstruction results using different FEM discretizations.

3 Influence of data noise

Here the influence of the uncertainty u resulting from the time-correlated single photon counting (TCSPC) is regarded [1]. The relative uncertainty u_{rel} in the maximum of the time-of-flight distributions of the poisson distributed measurement data can be controlled by the number of laser pulses for one measurement.

Noisy data are generated by adding a random vector function $\boldsymbol{p}$ to the simulation $\boldsymbol{J}^{mes} = \boldsymbol{J}^{sim}(\mu_a)+\boldsymbol{p}$. The components of $\boldsymbol{p}$ are Gauss-distributed with the mean value 0. Their standard deviations consider the correct uncertainty due to TCSPC experiments [1].

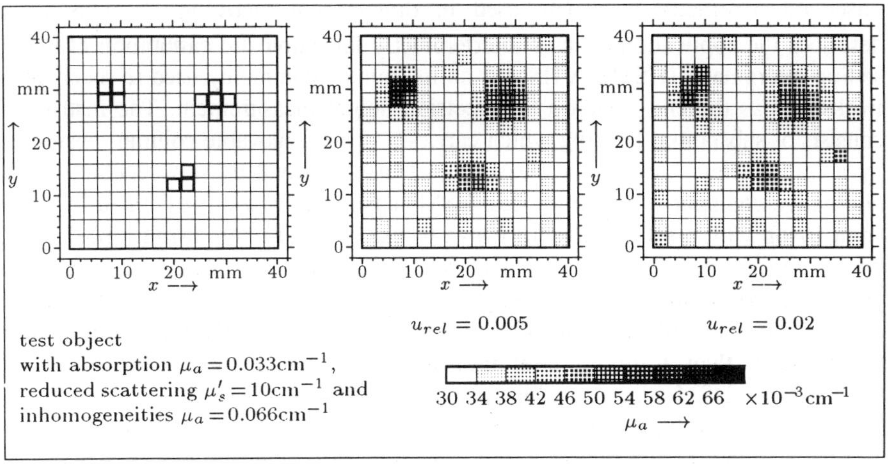

Figure 2: Influence of data noise on the reconstruction result.

Systematic tests show that the decrease of only one of both kinds of errors do not lead in all cases to better results. A good adjustment between both is necessary for an efficient algorithm.

1. D. V. O'Connor, D. Phillips, Time-correlated Single Photon Counting. Academic Press, London, Orlando, San Diego (1984).
2. A. Ishimaru, Wave propagation in random scattering media. New York (1978).
3. R. Model, R. Hünlich, M. Orlt, M. Walzel, Image reconstruction for random media by diffusion tomography. Proc. SPIE **2389**, 400–410 (1995).
4. M. Orlt, M. Walzel, R. Model, Transillumination imaging performance using time domain data, SPIE **2626**, 346–357 (1995).

Advanced Mathematical Tools in Metrology III
Edited by P. Ciarlini, M.G. Cox, F. Pavese & D. Richter

APPLICATION OF THE LASER INTERFEROMETRY FOR AUTOMATIC CALIBRATION OF HEIGHT SETTING MICROMETER

C. QUAN, G. XU
National Measurement Centre, Singapore Productivity and Standards Board,
1 Science Park Drive, Singapore Science Park, Singapore 118221

1 Introduction

Laser interferometry is widely used in dimensional metrology[1-4]. This paper describes the application of laser interferometry for automatic calibration of height setting micrometer (HSM). A HSM is an instrument commonly used in mechanical workshops as the end standard for precision height measurement. The instrument consists of substantial housing which carries a vertical movable column, positioned by a micrometer screw, with measuring elements provided with regular spaced, alternative coplanar measuring faces. HSMs have usually been calibrated by direct comparison with a set of standard gauge blocks or length bars using a positioning device and an electronic displacement sensor. Calibration of HSM in this way is very time consuming, typically requiring two or three days and also it needs constant attention of operator. The uncertainty for this method of calibration is hardly better than 1 μm. A laser measurement system reported here is used to calibrate the HSM automatically. The instrument measurement uncertainty for height of 350 mm is less than 0.5 μm at the 95% confidence level. The system can also be used to measure the other end standards, such as gauge blocks, length bars with high accuracy.

2 Description of the Measurement System

The system consists of double frequency laser interferometer (ZYGO AXIOM 2/20), a vertical air-bearing slide and intelligent digital axis controller, a X-Y positioning stage and DMC-600 PC Bus controller, and an electronic linear probe (TESA MODUL). The air-bearing mechanism guarantees smooth, friction-free movement of the slide and attached measuring elements and minimises motion errors such as run-out, vibration, and wavering line of travel. An interactive software tool has been developed for automatic calibration of HSM. The program is menu driven to control the vertical air bearing slide and horizontal X-Y movements. Laser and electronic probe readings are captured by IBM PC through RS232 interface and processed by IBM PC.

Since the unit of measure for laser measurement system is the wavelength of light, this quantity must be known at all the times. At the beginning of each calibration, the refractive index of the air, n_0, is calculated using Edlen's equation[5] by measuring air pressure, humidity and temperature. Once the initial value of refractive index of air is determined, the optical wavelength compensator tracks the changes in the refractive index of surrounding air from the initial value.

3. Uncertainty of Measurement

The uncertainty reached in HSM calibration is estimated by considering the known sources of error. The treatment of uncertainty here is in accordance with the "ISO Guide to the expression of the uncertainty in measurement[6]". The uncertainty budget distinguishes between contributions of Type A and Type B. The final system verification is based on comparison measurement with other length measurement instrument within the laboratory and the test certificate of commercial calibration laboratory.

The combined standard uncertainty is multiplied by a factor of k = 2.12 (corresponding to DoF = 16) to covert it to a 95% confidence level. The expanded measurement uncertainty is given by

$$U(H) = kU_c(H) = \pm(0.57H + 0.26)\mu m \qquad (1)$$

where H is the height of each elements of HSM in metres. Therefore, the best uncertainty at the height of 350 mm is 0.46 μm.

4. Results

The system is located in the laboratory at the well controlled environment (T = 20 ± 0.3°C, P = 1013 ± 3 mbar, H = 50 ± 5%). Testing consisted of measuring the Mitutoyo HSM which was calibrated before. In this measurement, the lower surface of HSM was measured for both columns. The results were calculated as the mean of three run measurements and well agreed with the previous calibration results. The next testing of the system was to compare the results with that of obtained from the NPL-TESA gauge block interferometer (primary standard) for length under 100 mm and Starrett (USA) Certificate of Calibration values for length bars from 125 mm to 304.8 mm. The maximum deviation of the comparison results is 271.6 nm. The estimated uncertainty of HSM and results of the comparison measurements are in good agreement with the gauge blocks and length bars up to 300 mm.

References

1. G. Xu and K. K. Koh, *NCSL Proceedings,* 9-15 (1994).
2. E. Ikinen and K. Riski, *Metrologia,* **30,** 95-104 (1993).
3. H. Darnedde, *Metrologia,* **29**, 349-359 (1992).
4. U. Brand and K. Herrmann, *Meas. Sci. Technol.* **7**, 911-917 (1996).
5. B. Edlen, 1966, *Metrologia,* **2(2),** 71(1966).
6. ISO/TAG4/WG3,*Guide to the expression of uncertainty in measurement,* 1993.

Author Index

Series on Advances in Mathematics for Applied Sciences

Series on Advances in Mathematics for Applied Sciences

Aims and Scope

This Series reports on new developments in mathematical research relating to methods, qualitative and numerical analysis, mathematical modeling in the applied and the technological sciences. Contributions related to constitutive theories, fluid dynamics, kinetic and transport theories, solid mechanics, system theory and mathematical methods for the applications are welcomed.

This Series includes books, lecture notes, proceedings, collections of research papers. Monograph collections on specialized topics of current interest are particularly encouraged. Both the proceedings and monograph collections will generally be edited by a Guest editor.

High quality, novelty of the content and potential for the applications to modern problems in applied science will be the guidelines for the selection of the content of this series.

Instructions for Authors

Submission of proposals should be addressed to the editors-in-charge or to any member of the editorial board. In the latter, the authors should also notify the proposal to one of the editors-in-charge. Acceptance of books and lecture notes will generally be based on the description of the general content and scope of the book or lecture notes as well as on sample of the parts judged to be more significantly by the authors.

Acceptance of proceedings will be based on relevance of the topics and of the lecturers contributing to the volume.

Acceptance of monograph collections will be based on relevance of the subject and of the authors contributing to the volume.

Authors are urged, in order to avoid re-typing, not to begin the final preparation of the text until they received the publisher's guidelines. They will receive from World Scientific the instructions for preparing camera-ready manuscript.

SERIES ON ADVANCES IN MATHEMATICS FOR APPLIED SCIENCES

Vol. 17 The Fokker-Planck Equation for Stochastic Dynamical Systems and Its Explicit Steady State Solution
by C. Soize

Vol. 18 Calculus of Variation, Homogenization and Continuum Mechanics
eds. G. Bouchitté et al.

Vol. 19 A Concise Guide to Semigroups and Evolution Equations
by A. Belleni-Morante

Vol. 20 Global Controllability and Stabilization of Nonlinear Systems
by S. Nikitin

Vol. 21 High Accuracy Non-Centered Compact Difference Schemes for Fluid Dynamics Applications
by A. I. Tolstykh

Vol. 22 Advances in Kinetic Theory and Computing: Selected Papers
ed. B. Perthame

Vol. 23 Waves and Stability in Continuous Media
eds. S. Rionero and T. Ruggeri

Vol. 24 Impulsive Differential Equations with a Small Parameter
by D. Bainov and V. Covachev

Vol. 25 Mathematical Models and Methods of Localized Interaction Theory
by A. I. Bunimovich and A. V. Dubinskii

Vol. 26 Recent Advances in Elasticity, Viscoelasticity and Inelasticity
ed. K. R. Rajagopal

Vol. 27 Nonstandard Methods for Stochastic Fluid Mechanics
by M. Capinski and N. J. Cutland

Vol. 28 Impulsive Differential Equations: Asymptotic Properties of the Solutions
by D. Bainov and P. Simeonov

Vol. 29 The Method of Maximum Entropy
by H. Gzyl

Vol. 30 Lectures on Probability and Second Order Random Fields
by D. B. Hernández

Vol. 31 Parallel and Distributed Signal and Image Integration Problems
eds. R. N. Madan et al.

Vol. 32 On the Way to Understanding The Time Phenomenon: The Constructions of Time in Natural Science – Part 1. Interdisciplinary Time Studies
ed. A. P. Levich

SERIES ON ADVANCES IN MATHEMATICS FOR APPLIED SCIENCES

Vol. 33 Lecture Notes on the Mathematical Theory of the Boltzmann Equation
ed. N. Bellomo

Vol. 34 Singularly Perturbed Evolution Equations with Applications to Kinetic Theory
by J. R. Mika and J. Banasiak

Vol. 35 Mechanics of Mixtures
by K. R. Rajagopal and L. Tao

Vol. 36 Dynamical Mechanical Systems Under Random Impulses
by R. Iwankiewicz

Vol. 37 Oscillations in Planar Dynamic Systems
by R. E. Mickens

Vol. 38 Mathematical Problems in Elasticity
ed. R. Russo

Vol. 39 On the Way to Understanding the Time Phenomenon: The Constructions of Time in Natural Science — Part 2. The "Active" Properties of Time According to N. A. Kozyrev
ed. A. P. Levich

Vol. 40 Advanced Mathematical Tools in Metrology II
eds. P. Ciarlini et al.

Vol. 41 Mathematical Methods in Electromagnetism Linear Theory and Applications
by M. Cessenat

Vol. 42 Constrained Control Problems of Discrete Processes
by V. N. Phat

Vol. 43 Motor Vehicle Dynamics: Modeling and Simulation
by G. Genta

Vol. 44 Microscopic Theory of Condensation in Gases and Plasma
by A. L. Itkin and E. G. Kolesnichenko

Vol. 45 Advanced Mathematical Tools in Metrology III
eds. P. Cianlini et al.